影 印 版 说 明

MOMENTUM PRESS 出版的 *Plastics Technology Handbook*（2 卷）是介绍塑料知识与技术的大型综合性手册，内容涵盖了从高分子基本原理，到塑料的合成、种类、性能、配料、加工、制品，以及模具、二次加工等各个方面。通过阅读、学习本手册，无论是专业人员还是非专业人员，都会很快熟悉和掌握塑料制品的设计和制造方法。可以说一册在手，别无他求。

原版 2 卷影印时分为 11 册，第 1 卷分为：

  塑料基础知识·塑料性能

  塑料制品生产

  注射成型

  挤压成型

  吹塑成型

  热成型

  发泡成型·压延成型

第 2 卷分为：

  涂层·浇注成型·反应注射成型·旋转成型

  压缩成型·增强塑料·其他工艺

  模具

  辅机与二次加工设备

唐纳德 V·罗萨多，波士顿大学化学学士学位，美国东北大学 MBA 学位，马萨诸塞大学洛厄尔分校工程塑料和加州大学工商管理博士学位（伯克利）。著有诸多论文及著作，包括《塑料简明百科全书》、《注塑手册（第三版）》以及塑料产品材料和工艺选择手册等。活跃于塑料界几十年，现任著名的 Plasti Source Inc. 公司总裁，并是美国塑料工业协会（SPI）、美国塑料学会（PIA）和 SAMPE（The Society for the Advancement of Material and Process Engineering）的重要成员。

材料科学与工程图书工作室

  联系电话 0451-86412421

      0451-86414559

  邮 箱 yh_bj@aliyun.com

      xuyaying81823@gmail.com

      zhxh6414559@aliyun.com

影印版

PLASTICS TECHNOLOGY HANDBOOK

塑料技术手册

VOLUME 2

COMPRESSION MOLDING · REINFORCED PLASTIC · OTHER PROCESSES

压缩成型 · 增强塑料 · 其他工艺

EDITED BY

DONALD V. ROSATO
MARLENE G. ROSATO
NICK R. SCHOTT

哈尔滨工业大学出版社
HARBIN INSTITUTE OF TECHNOLOGY PRESS

黑版贸审字08-2014-093号

Donald V.Rosato, Marlene G.Rosato, Nick R.Schott
Plastics Technology Handbook Volume 2
9781606500828
Copyright © 2012 by Momentum Press, LLC
All rights reserved.

Originally published by Momentum Press, LLC
English reprint rights arranged with Momentum Press, LLC through McGraw-Hill Education (Asia)

This edition is authorized for sale in the People's Republic of China only, excluding Hong Kong, Macao SAR and Taiwan.

本书封面贴有McGraw-Hill Education公司防伪标签，无标签者不得销售。
版权所有，侵权必究。

图书在版编目（CIP）数据

塑料技术手册. 第2卷. 压缩成型·增强塑料·其他工艺 =Plastics technology handbook volume 2 compression molding·reinforced plastic·other processes : 英文 /（美）罗萨多（Rosato, D. V.）等主编. —影印本. — 哈尔滨：哈尔滨工业大学出版社，2015.6
　　ISBN 978-7-5603-5048-6

　　Ⅰ.①塑… Ⅱ.①罗… Ⅲ.①塑料–技术手册–英文 ②塑料成型–技术手册–英文 ③增强塑料–技术手册–英文 ④塑料–生产工艺–技术手册–英文 Ⅳ.①TQ320.6-62

中国版本图书馆CIP数据核字（2014）第280096号

责任编辑	张秀华　杨　桦　许雅莹
出版发行	哈尔滨工业大学出版社
社　　址	哈尔滨市南岗区复华四道街10号　邮编150006
传　　真	0451-86414749
网　　址	http://hitpress.hit.edu.cn
印　　刷	哈尔滨市石桥印务有限公司
开　　本	787mm×960mm　1/16　印张14
版　　次	2015年6月第1版　2015年6月第1次印刷
书　　号	ISBN 978-7-5603-5048-6
定　　价	70.00元

（如因印刷质量问题影响阅读，我社负责调换）

PLASTICS TECHNOLOGY HANDBOOK

VOLUME 2

EDITED BY

Donald V. Rosato, PhD, MBA, MS, BS, PE
PlastiSource Inc.
Society of Plastics Engineers
Plastics Pioneers Association
UMASS Lowell Plastics Advisory Board

Marlene G. Rosato, BASc (ChE), P Eng
Gander International Inc.
Canadian Society of Chemical Engineers
Association of Professional Engineers of Ontario
Product Development and Management Association

Nick R. Schott, PhD, MS, BS (ChE), PE
UMASS Lowell Professor of Plastics Engineering Emeritus & Plastics Department Head Retired
Plastics Institute of America
Secretary & Director for Educational and Research Programs

Momentum Press, LLC, New York

Contents

FIGURES	10
TABLES	14
ABBREVIATIONS	18
ACKNOWLEDGMENTS	26
PREFACE	27
ABOUT THE AUTHORS	30
14. COMPRESSION MOLDING	**178**
INTRODUCTION	178
MOLD	185
MACHINES	188
PLASTIC	193
Polytetrafluoroethylene Billet	196
Hot Compression-Molding PTFE	203
PROCESSING	204
Heating	205
Automation	207
Transfer Molding	211
Compression-Injection Molding	212
Compression and Isostatic Molding	216

15. REINFORCED PLASTIC — 223
OVERVIEW — 223
DEFINITION — 225
- Fibrous Composite — 240
- Laminar Composite — 251
- Particulate Composites — 252
- Fillers — 252
PROPERTIES — 254
ORIENTATION OF REINFORCEMENT — 270
- Directional Property — 274
- Hetergeneous/Homogeneous/Anisotropic — 279
MATERIAL OF CONSTRUCTION — 279
- Prepreg — 282
- Sheet Molding Compound — 283
- Bulk Molding Compound — 284
- Compound — 285
FABRICATING PROCESS — 286
- Preform Process — 286
- Type Process — 288
- Compression Molding — 288
- Hand Layup — 291
- Filament Winding — 295
- Injection Molding — 306
- Marco Process — 307
- Pultrusion — 307
- Reactive Liquid Molding — 309
- Reinforced RTM — 310
- Reinforced Rotational Molding — 311
- SCRIMP Process — 311
- Soluble Core Molding — 312
- Spray-Up — 312
- Stamping — 314
SELECTING PROCESSES — 315
DESIGN — 317
- Aspect Ratio — 317
- Tolerance — 329
ENGINEERING ANALYSIS — 333
- Design Theory — 333

16. OTHER PROCESSES — 335
- INTRODUCTION — 335
- PVC PLASTISOL — 336
  - Introduction — 336
  - Processing Plastisol — 338
  - Processing Organosol — 340
  - Slush Molding — 340
  - Rotational Molding — 341
  - Spray Molding — 342
  - Continuous Coating — 342
  - Open Molding — 342
  - Closed Molding — 343
  - Dip Molding — 343
  - Dip Coating — 344
  - Heating System — 344
- INK SCREENING — 344
- ENCAPSULATION — 344
- POTTING — 345
- LIQUID INJECTION MOLDING — 345
  - Vacuum-Assisted LIM — 346
- IMPREGNATION — 346
- CHEMICAL ETCHING — 347
- TWIN-SCREW INJECTION MOLDING — 347
- TEXTILE COVERED MOLDING — 348
- MELT COMPRESSION MOLDING — 348
  - Back Injection — 349
  - Melt Flow Compression Molding — 351
  - Back Compression (Melt Compression Molding) — 352
- MCM-IML — 352
- PROCESSING COMPARISON — 353

Figures

Figure 14.1	Schematic of the CM of a plastic material.	178
Figure 14.2	Compression molded ring-shaped part removed from the mold.	179
Figure 14.3	CM using a molding compound.	182
Figure 14.4	CM using an impregnated material.	182
Figure 14.5	Examples of flash in a mold: (a) horizontal, (b) vertical, and (c) modified vertical.	184
Figure 14.6	Positive compression mold.	186
Figure 14.7	Flash compression mold.	186
Figure 14.8	Semipositive compression mold.	187
Figure 14.9	Example of mold vent locations.	187
Figure 14.10	Example of vent locations in a mold processing TPs.	188
Figure 14.11	Example of land locations in a split-wedge mold (courtesy of National Tool and Manufacturing Association).	189
Figure 14.13	The left side is a better edge design when using a draw angle.	190
Figure 14.12	Optimum draft for shear edges in molding sheet-molding compounds.	190
Figure 14.14	Knife shear edge.	190
Figure 14.15	Press with 4 × 4 in platens and ½-ton clamp pressure (courtesy of Carver Press).	191
Figure 14.16	A 400-ton press with much larger than normal platens that measure 5 × 10 ft; the press has multiple zones of electrically heated platens, an automatic bump cycle, an audible alarm to signal the end of the cure cycle, and front and back safety-light curtains (courtesy of Wabash MPI).	191

Figure 14.17	A 4000-ton press with 5 × 8 ft platens (courtesy of Erie Press).	192
Figure 14.18	A 400-ton press with 18 platens, each measuring 4 × 6 ft (courtesy of Baldwin Works).	193
Figure 14.19	An 8000-ton press with 10 × 10 ft platens that have book-type opening and closing action (courtesy of Krismer, Germany).	194
Figure 14.20	Processing sequence for compression stamping glass fiber–reinforced TP sheets.	195
Figure 14.21	Heat-curing cycles for TPs go through A-B-C stages.	195
Figure 14.22	Transition point and linear thermal expansion of PTFE (courtesy of DuPont).	199
Figure 14.23	Mechanism of sintering PTFE (courtesy of DuPont).	200
Figure 14.24	Example of a sintering cycle.	202
Figure 14.25	Example of a simple loading tray with a retractable slide plate to deliver material to multicavity mold.	207
Figure 14.26	CM machine with preplasticizer.	208
Figure 14.27	Three screws of the preplasticizer have been retracted from their barrels for viewing; not in the operating mode.	209
Figure 14.28	Preheated compounds exiting the preplasticizers prior to guillotine slicing the required shot sizes.	210
Figure 14.29	Schematic of transfer molding.	211
Figure 14.30	Comparing IM, CM, and transfer molding.	211
Figure 14.31	Detail view of transfer molding with two cavities.	212
Figure 14.32	Example of a screw plasticizer preheating plastic that is delivered into the transfer molding pot for delivery into the mold cavities.	212
Figure 14.33	A 64-cavity transfer mold about to receive electronic devices from a work-loading frame.	215
Figure 14.34	Principal steps of isostatic molding.	217
Figure 14.35	Basic isostatic compaction process.	219
Figure 14.36	Three ways of molding PTFE tubes: (a) two flexible bags, (b) inner flexible bag with outer rigid cylinder, and (c) outer flexible bag with inner rigid rod.	220
Figure 15.1	Effect of matrix content on strength (F) or elastic moduli (E) of RPs.	223
Figure 15.2	Properties versus amount of reinforcement.	224
Figure 15.3	Glass fiber-TS polyester-filament-wound RP underground gasoline storage tank.	226
Figure 15.4	Complete primary and secondary bus structure hand layup of glass fiber-TS polyester RP.	226
Figure 15.5	Glass fiber swirl mat-TS polyester RP vacuum hand layup boat shell.	227
Figure 15.6	Glass fiber-TS polyester RP robot controlled hand layup 28 ft long boat.	227

Figure 15.8	Glass fiber-TS polyester filament wound RP tank trailer that transports corrosive and hazardous materials.	228
Figure 15.7	Glass fiber tape-TS polyester hand layup smoke stack liner.	228
Figure 15.9	Pultruded glass fiber roving-TS polyester rods in a 370 ft long lift bridge supports up to 44 T traffic load.	228
Figure 15.10	Glass fiber-TS polyester filament wound RP railroad hopper car body.	229
Figure 15.11	Monsanto House of the future all glass fiber-TS polyester RP hand layup has four 16 ft long U-shaped (monocoque box girders) cantilever structures 90° apart producing the main interior.	229
Figure 15.12	Interface of a RP.	230
Figure 15.13	Examples of reinforcement types and processing methods.	230
Figure 15.14	Fishbone diagram for an RP process (courtesy of Plastics FALLO).	231
Figure 15.15	Review of different processes to fabricate RP products.	231
Figure 15.16	Modulus of different materials can be related to their specific gravities with RPs providing an interesting graph.	232
Figure 15.17	Short and long glass fiber-TP RP data (wt% fiber in parentheses).	246
Figure 15.18	Short to long fibers influence properties of RPs.	247
Figure 15.19	Specific tensile strength to specific tensile modulus of elasticity data f nylon RPs.	247
Figure 15.20	Flexural fatigue data of woven glass fiber roving RPs.	247
Figure 15.21	Common glass fiber-TS polyester resin RP fatigue data versus other materials (chapter 19).	248
Figure 15.22	Comparing different fiber material strength properties at elevated temperatures.	248
Figure 15.23	Comparing whisker reinforcements with other reinforcements.	249
Figure 15.24	Schematic example in the manufacture of glass filaments/fibers.	249
Figure 15.25	Staple glass fiber and continuous glass filament fiber process methods.	272
Figure 15.26	Fiber arrangements and property behavior (courtesy of Plastics FALLO).	272
Figure 15.27	RP density versus percentage glass by weight or volume.	273
Figure 15.28	Fiber orientation provides different directional properties.	274
Figure 15.29	Examples of how fiber orientation influences properties of RPs.	275
Figure 15.30	Parallel/bidirectional layup of woven fabric 181 glass fiber (courtesy of Plastics FALLO).	280
Figure 15.31	Parallel/unidirectional layup woven fabric 143 glass fiber (courtesy of Plastics FALLO).	280
Figure 15.32	Ply layup at 0° and 90° woven fabric 143 glass fiber construction (courtesy of Plastics FALLO).	281
Figure 15.33	Ply layup at 0°, 45°, 90°, and 135° woven fabric 143 glass fiber construction (courtesy of Plastics FALLO).	281

Figure 15.34	Sheet molding compound (SMC) production line using chopped glass fiber including roving to provide bidirectional properties, cutting continuous rovings for ease of mold-cavity fit.	282
Figure 15.35	These different SMC production lines produce by using chopped glass fibers (top), including roving to provide bidirectional properties, cutting continuous rovings so that they can fit easily in a mold cavity, and producing thicker SMC (about 4 mm thick by 120 cm wide; bottom).	284
Figure 15.36	Flow of glass fiber rovings traveling through a plenum machine.	287
Figure 15.38	Flow of glass fiber rovings traveling through a water-slurry machine.	287
Figure 15.37	Flow of glass fiber rovings traveling through a direct machine.	287
Figure 15.39	Two-part compression mold.	289
Figure 15.40	Layout of reinforcement is designed to meet structural requirements.	293
Figure 15.41	Automated-integrated RP vacuum hand layup process that uses prepreg sheets that are in the B-stage (chapter 1).	293
Figure 15.42	Schematic of hand-layup bag molding in an autoclave.	294
Figure 15.43	Early-twentieth-century tape-wrapping patent of a tube-making machine by Hoganas-Billesholms A.B., Sweden.	297
Figure 15.44	Views of fiber filament-wound isotensoid pattern of the reinforcing fibers without plastic (left) and with resin cured.	301
Figure 15.45	Box winding machine with position changes of clamp tooling.	301
Figure 15.46	Schematics of "racetrack" filament-winding machines. Top view shows machine in action; other view is a schematic of a machine built to fabricate 150,000 gal rocket motor tanks.	304
Figure 15.47	Conventional single stage IMM.	306
Figure 15.48	IM with a preloader usually providing heat to the RP compound.	307
Figure 15.49	Schematics of ram and screw IMM.	308
Figure 15.50	Use is made of vacuum, pressure, or pressure-vacuum in the Marco process.	309
Figure 15.51	Cutaway view of a reinforced RTM mold.	311
Figure 15.52	Lost-wax process fabricated a high-strength RP structural beam.	312
Figure 15.53	Nonatomized, dispensed Glass-Craft spray gun is easy to use and produces low styrene emissions and is economic to maintain.	313
Figure 15.54	Example of the effect of shrinkage in the longitudinal and transverse directions of a molded part.	319
Figure 15.55	Tensile stress-strain curves for epoxy-unreinforced and epoxy-reinforced RPs and other materials.	322
Figure 15.56	Example of crack propagation to fracture that can occur, resulting in product failure under load.	329
Figure 16.1	Effect of temperature on macromolecular characteristics of PVC plastisol.	337
Figure 16.2	Example of time-dependent viscosity of PVC plastisol.	338

Tables

Table 14.1	Example of applications for compression molded thermoset (TS) plastics	180
Table 14.2	Comparing compression molded properties with other processes	180
Table 14.3	Relating materials to properties to processes	181
Table 14.4	Examples of the effect of preheating and part depth of phenolic parts on CM pressure (psi)	183
Table 14.5	Examples of OD, ID, height, and weight relationships of different PTFE billet CMs	197
Table 14.6	Examples of PTFE sintering conditions	201
Table 14.7	Effect of cooling rate on crystallinity, typical for granular molding powders (courtesy of DuPont)	202
Table 14.8	Effect of CM processes on properties (courtesy of DuPont)	204
Table 14.9	Guide to wall-thickness tolerance for CM different plastics	205
Table 14.10	Guide in the use of reinforcements and fillers in different molding compounds	206
Table 14.11	Transfer molding compared to CM	213
Table 14.12	Transfer molding compared to reinforced plastic molding	214
Table 14.13	Examples of isostatically molded parts	217
Table 14.14	Isostatic mold design considerations	222
Table 15.1	Types of composites	224
Table 15.2	Examples of composite ablative compounds	224
Table 15.3	Examples of reinforcement types and processing methods	232
Table 15.4	Examples of RTP properties	233
Table 15.5	TP-glass fiber RPs injection molding (IM) temperatures	234
Table 15.6	Examples of properties and processes of RTS plastics	235
Table 15.7	Properties of the popular TS polyester-glass fiber RPs	235

Table 15.8	Different properties of RTPs and RTSs per ASTM standards	236
Table 15.9	Properties of fiber reinforcements	240
Table 15.10	Reinforcement thermal properties	240
Table 15.11	Properties of glass-fiber RPs	241
Table 15.12	Comparative yarn properties	242
Table 15.13	Examples of different carbon fibers	242
Table 15.14	Aramid fiber-TP RP properties	242
Table 15.15	Properties of unidirectional hybrid-nylon RPs	243
Table 15.16	Charpy impact test results of square woven fabric using hybrid fibers-nylon RPs	244
Table 15.17	Damage propagation of aramid and E-glass RPs using tensile-notched test specimens	244
Table 15.18	Examples of different glass fiber yarns	244
Table 15.19	Examples of glass fiber staple fiber yarn data	245
Table 15.20	Examples of glass fiber cloth constructions	246
Table 15.21	Examples of fillers used in TP RPs (chapter 1)	253
Table 15.22	Examples of fillers used in TS RPs (chapter 1)	253
Table 15.23	Comparison of tensile properties in RPs, steel, and aluminum	254
Table 15.24	Mechanical properties of resins that are reinforced to increase properties	255
Table 15.25	Properties per ASTM of 30 wt% glass-fiber RTPs	256
Table 15.26	Properties of glass-fiber RTPs with different glass fiber contents and other reinforcements	257
Table 15.27	Properties of short and long glass fiber-nylon 6/6 RPs at elevated temperatures	257
Table 15.28	Examples of obtaining desired properties of TP-RPs	258
Table 15.29	Properties of RPs with 30 wt% to 50 wt% glass fiber-TS polyester based on fabricating process	259
Table 15.30	Properties of TS polyester RPs with different amounts of glass fibers	260
Table 15.31	Properties of glass fiber mats RPs with different types of TS polyesters	261
Table 15.32	General properties of TS RPs per ASTM testing procedures	262
Table 15.33	Examples of mechanical properties of TS RPs at ambient and elevated temperatures	264
Table 15.34	Flexural modulus of glass-polyester–RPs exposed to various environmental elements	265
Table 15.35	Strength and modulus for glass fiber-TS RPs at low temperature	266
Table 15.36	Coefficients of thermal expansion for parallel glass fiber-TS RPs	267
Table 15.37	Example of TS RPs for electrical applications	268
Table 15.38	Mechanical properties of glass fabric-TS polyester RPs exposed to various intensities of near-UV radiation in a vacuum	269
Table 15.39	Mechanical properties of glass fiber fabric-TS polyester RPs after irradiation at elevated temperatures	270

Table 15.40	Properties of different materials	271
Table 15.41	Properties of unidirectional RPs using different types of fibers	276
Table 15.42	Properties of unidirectional graphite fiber-thermoplastic RPs varying in resin content by weight and varying in void content by volume (at 72°F and 350°F)	277
Table 15.43	Comparing properties of SMC with steel	283
Table 15.44	Filament-wound structures for commercial and industrial applications	296
Table 15.45	Filament-wound structures for aerospace, hydrospace, and military applications	297
Table 15.46	Different FW patterns meet different performance requirements	298
Table 15.47	RP processing guide to RP process selection	316
Table 15.48	RP processing guide to RP size	317
Table 15.49	Examples of a few processes to material comparisons	318
Table 15.50	RP resin transfer, SMC compression, and IM processes compared	319
Table 15.51	Examples of RTS plastic processes	320
Table 15.52	Comparing uses of different plastics with different RP and other processes	321
Table 15.53	Examples of interrelating product-RP material-process performances	322
Table 15.54	Comparison of RP design aspects and processes to cost	323
Table 15.55	Examples of processing variables	325
Table 15.56	Product design versus processing methods	326
Table 15.57	Other product design considerations versus processing methods	327
Table 15.58	Product design shapes versus processing methods	328
Table 15.59	Examples of the efficiency RPs fiber orientation	329
Table 15.60	Example of TS polyester volume shrinkage during curing	330
Table 15.61	RPs wall-thickness tolerances	331
Table 15.62	Comparing unreinforced and RP mold shrinkage rates	332
Table 15.63	Composite efficiency of RPs	334
Table 15.64	Examples of loading conditions	334
Table 16.1	Example of a PVC blend formulation	343
Table 16.2	Automotive industry objectives for decorative plastics	349
Table 16.3	Definitions applicable to low-pressure decorating molding	350
Table 16.4	Example of an MCM-IML molding cycle	352
Table 16.5	Examples of MCM-IML advantages and applications	353
Table 16.6	Examples of valid reasons for using MCM-IML	354
Table 16.7	Examples of invalid reasons for using MCM-IML	354
Table 16.8	Process and materials composition	355
Table 16.9	Processing, materials, and geometry	355
Table 16.10	Geometry function and complexity	356
Table 16.11	Listing of abbreviations used in the following tables	357
Table 16.12	Reactive liquid composite molding	358

Table 16.13	Multimaterial multiprocess (MMP) technology	359
Table 16.15	TP sheet composite	360
Table 16.14	Fusible core IM	360
Table 16.16	Gas-assisted IM: process and simulation	361
Table 16.17	Low-pressure molding	362
Table 16.18	Advanced blow molding	363
Table 16.19	Microcellular plastic: formation and shaping	364
Table 16.20	Lamellar IM	365

Abbreviations

AA acrylic acid
AAE American Association of Engineers
AAES American Association of Engineering Societies
ABR polyacrylate
ABS acrylontrile-butadiene-styrene
AC alternating current
ACS American Chemical Society
ACTC Advanced Composite Technology Consortium
ad adhesive
ADC allyl diglycol carbonate (also CR-39)
AFCMA Aluminum Foil Container Manufacturers' Association
AFMA American Furniture Manufacturers' Association
AFML Air Force Material Laboratory
AFPA American Forest and Paper Association
AFPR Association of Foam Packaging Recyclers
AGMA American Gear Manufacturers' Association
AIAA American Institute of Aeronautics and Astronauts
AIChE American Institute of Chemical Engineers
AIMCAL Association of Industrial Metallizers, Coaters, and Laminators
AISI American Iron and Steel Institute
AMBA American Mold Builders Association
AMC alkyd molding compound
AN acrylonitrile
ANSI American National Standards Institute
ANTEC Annual Technical Conference (of the Society of the Plastic Engineers)
APC American Plastics Council
APET amorphous polyethylene terephthalate
APF Association of Plastics Fabricators
API American Paper Institute
APME Association of Plastics Manufacturers in Europe
APPR Association of Post-Consumer Plastics Recyclers
AQL acceptable quality level
AR aramid fiber; aspect ratio
ARP advanced reinforced plastic
ASA acrylonitrile-styrene-acrylate
ASCII american standard code for information exchange
ASM American Society for Metals

ASME American Society of Mechanical Engineers
ASNDT American Society for Non-Destructive Testing
ASQC American Society for Quality Control
ASTM American Society for Testing Materials
atm atmosphere
bbl barrel
BFRL Building and Fire Research Laboratory
Bhn Brinell hardness number
BM blow molding
BMC bulk molding compound
BO biaxially oriented
BOPP biaxially oriented polypropylene
BR polybutadiene
Btu British thermal unit
buna polybutadiene
butyl butyl rubber
CA cellulose acetate
CAB cellulose acetate butyrate
CaCO$_3$ calcium carbonate (lime)
CAD computer-aided design
CAE computer-aided engineering
CAM computer-aided manufacturing
CAMPUS computer-aided material preselection by uniform standards
CAN cellulose acetate nitrate
CAP cellulose acetate propionate
CAS Chemical Abstract Service (a division of the American Chemical Society)
CAT computer-aided testing
CBA chemical blowing agent
CCA cellular cellulose acetate
CCV Chrysler composites vehicle
CEM Consorzio Export Mouldex (Italian)
CFA Composites Fabricators Association
CFC chlorofluorocarbon
CFE polychlorotrifluoroethylene
CIM ceramic injection molding; computer integrated manufacturing
CLTE coefficient of linear thermal expansion
CM compression molding
CMA Chemical Manufacturers' Association
CMRA Chemical Marketing Research Association
CN cellulose nitrate (celluloid)
CNC computer numerically controlled
CP Canadian Plastics
CPE chlorinated polyethylene
CPET crystallized polyethylene terephthalate
CPI Canadian Plastics Institute
cpm cycles/minute
CPVC chlorinated polyvinyl chloride
CR chloroprene rubber; compression ratio
CR-39 allyl diglycol carbonate
CRP carbon reinforced plastics
CRT cathode ray tube
CSM chlorosulfonyl polyethylene
CTFE chlorotrifluorethylene
DAP diallyl phthalate
dB decibel
DC direct current
DEHP diethylhexyl phthalate
den denier
DGA differential gravimetric analysis
DINP diisononyl phthalate
DMA dynamic mechanical analysis
DMC dough molding compound
DN *Design News* publication
DOE Design of Experments
DSC differential scanning calorimeter
DSD Duales System Deutschland (German Recycling System)
DSQ German Society for Quality
DTA differential thermal analysis
DTGA differential thermogravimetric analysis
DTMA dynamic thermomechanical analysis
DTUL deflection temperature under load
DV devolatilization
DVR design value resource; dimensional velocity research; Druckverformungsrest (German

compression set); dynamic value research; dynamic velocity ratio
E modulus of elasticity; Young's modulus
EBM extrusion blow molding
E_c modulus, creep (apparent)
EC ethyl cellulose
ECTFE polyethylene-chlorotrifluoroethylene
EDM electrical discharge machining
E/E electronic/electrical
EEC European Economic Community
EI modulus × moment of inertia (equals stiffness)
EMI electromagnetic interference
EO ethylene oxide (also EtO)
EOT ethylene ether polysulfide
EP ethylene-propylene
EPA Environmental Protection Agency
EPDM ethylene-propylene diene monomer
EPM ethylene-propylene fluorinated
EPP expandable polypropylene
EPR ethylene-propylene rubber
EPS expandable polystyrene
E_r modulus, relaxation
E_s modulus, secant
ESC environmental stress cracking
ESCR environmental stress cracking resistance
ESD electrostatic safe discharge
ET ethylene polysulfide
ETFE ethylene terafluoroethylene
ETO ethylene oxide
EU entropy unit; European Union
EUPC European Association of Plastics Converters
EUPE European Union of Packaging and Environment
EUROMAP Eu^ropean Committee of Machine Manufacturers for the Rubber and Plastics Industries (Zurich, Switzerland)
EVA ethylene-vinyl acetate
E/VAC ethylene/vinyl acetate copolymer
EVAL ethylene-vinyl alcohol copolymer (tradename for EVOH)
EVE ethylene-vinyl ether
EVOH ethylene-vinyl alcohol copolymer (or EVAL)
EX extrusion
F coefficient of friction; Farad; force
FALLO follow all opportunities
FDA Food and Drug Administration
FEA finite element analysis
FEP fluorinated ethylene-propylene
FFS form, fill, and seal
FLC fuzzy logic control
FMCT fusible metal core technology
FPC flexible printed circuit
fpm feet per minute
FRCA Fire Retardant Chemicals Association
FRP fiber reinforced plastic
FRTP fiber reinforced thermoplastic
FRTS fiber reinforced thermoset
FS fluorosilicone
FTIR Fourier transformation infrared
FV frictional force × velocity
G gravity; shear modulus (modulus of rigidity); torsional modulus
GAIM gas-assisted injection molding
gal gallon
GB gigabyte (billion bytes)
GD&T geometric dimensioning and tolerancing
GDP gross domestic product
GFRP glass fiber reinforced plastic
GMP good manufacturing practice
GNP gross national product
GP general purpose
GPa giga-Pascal
GPC gel permeation chromatography
gpd grams per denier
gpm gallons per minute
GPPS general purpose polystyrene
GRP glass reinforced plastic
GR-S polybutadiene-styrene
GSC gas solid chromatography

H hysteresis; hydrogen
HA hydroxyapatite
HAF high-abrasion furnace
HB Brinell hardness number
HCFC hydrochlorofluorocarbon
HCl hydrogen chloride
HDPE high-density polyethylene (also PE-HD)
HDT heat deflection temperature
HIPS high-impact polystyrene
HMC high-strength molding compound
HMW-HDPE high molecular weight–high density polyethylene
H-P Hagen-Poiseuille
HPLC high-pressure liquid chromatography
HPM hot pressure molding
HTS high-temperature superconductor
Hz Hertz (cycles)
I integral; moment of inertia
IB isobutylene
IBC internal bubble cooling
IBM injection blow molding; International Business Machines
IC *Industrial Computing* publication
ICM injection-compression molding
ID internal diameter
IEC International Electrochemical Commission
IEEE Institute of Electrical and Electronics Engineers
IGA isothermal gravimetric analysis
IGC inverse gas chromatography
IIE Institute of Industrial Engineers
IM injection molding
IMM injection molding machine
IMPS impact polystyrene
I/O input/output
ipm inch per minute
ips inch per second
IR synthetic polyisoprene (synthetic natural rubber)
ISA Instrumentation, Systems, and Automation
ISO International Standardization Organization or International Organization for Standardization
IT information technology
IUPAC International Union of Pure and Applied Chemistry
IV intrinsic viscosity
IVD in vitro diagnostic
J joule
JIS Japanese Industrial Standard
JIT just-in-time
JIT just-in-tolerance
J_p polar moment of inertia
JSR Japanese SBR
JSW Japan Steel Works
JUSE Japanese Union of Science and Engineering
JWTE Japan Weathering Test Center
K bulk modulus of elasticity; coefficient of thermal conductivity; Kelvin; Kunststoffe (plastic in German)
kb kilobyte (1000 bytes)
kc kilocycle
kg kilogram
KISS keep it short and simple
Km kilometer
kPa kilo-Pascal
ksi thousand pounds per square inch (psi $\times 10^3$)
lbf pound-force
LC liquid chromatography
LCP liquid crystal polymer
L/D length-to-diameter (ratio)
LDPE low-density polyethylene (PE-LD)
LIM liquid impingement molding; liquid injection molding
LLDPE linear low-density polyethylene (also PE-LLD)
LMDPE linear medium density polyethylene
LOX liquid oxygen
LPM low-pressure molding
m matrix; metallocene (catalyst); meter

mμ micromillimeter; millicron; 0.000001 mm
μm micrometer
MA maleic anhydride
MAD mean absolute deviation; molding area diagram
Mb bending moment
MBTS benzothiazyl disulfide
MD machine direction; mean deviation
MD&DI Medical Device and Diagnostic Industry
MDI methane diisocyanate
MDPE medium density polyethylene
Me metallocene catalyst
MF melamine formaldehyde
MFI melt flow index
mHDPE metallocene high-density polyethylene
MI melt index
MIM metal powder injection molding
MIPS medium impact polystyrene
MIT Massachusetts Institute of Technology
mLLDPE metallocene catalyst linear low-density polyethylene
MMP multimaterial molding or multimaterial multiprocess
MPa mega-Pascal
MRPMA Malaysian Rubber Products Manufacturers' Association
Msi million pounds per square inch (psi × 10^6)
MSW municipal solid waste
MVD molding volume diagram
MVT moisture vapor transmission
MW molecular weight
MWD molecular weight distribution
MWR molding with rotation
N Newton (force)
NACE National Association of Corrosion Engineers
NACO National Association of CAD/CAM Operation
NAGS North America Geosynthetics Society
NASA National Aeronautics Space Administration
NBR butadiene acrylontrile
NBS National Bureau of Standards (since 1980 renamed the National Institute Standards and Technology or NIST)
NC numerical control
NCP National Certification in Plastics
NDE nondestructive evaluation
NDI nondestructive inspection
NDT nondestructive testing
NEAT nothing else added to it
NEMA National Electrical Manufacturers' Association
NEN Dutch standard
NFPA National Fire Protection Association
NISO National Information Standards Organization
NIST National Institute of Standards and Technology
nm nanometer
NOS not otherwise specified
NPCM National Plastics Center and Museum
NPE National Plastics Exhibition
NPFC National Publications and Forms Center (US government)
NR natural rubber (polyisoprene)
NSC National Safety Council
NTMA National Tool and Machining Association
NWPCA National Wooden Pallet and Container Association
OD outside diameter
OEM original equipment manufacturer
OPET oriented polyethylene terephthalate
OPS oriented polystyrene
OSHA Occupational Safety and Health Administration
P load; poise; pressure
Pa Pascal
PA polyamide (nylon)
PAI polyamide-imide
PAN polyacrylonitrile

PB polybutylene
PBA physical blowing agent
PBNA phenyl-β-naphthylamine
PBT polybutylene terephthalate
PC permeability coefficient; personal computer; plastic composite; plastic compounding; plastic-concrete; polycarbonate; printed circuit; process control; programmable circuit; programmable controller
PCB printed circuit board
pcf pounds per cubic foot
PCFC polychlorofluorocarbon
PDFM Plastics Distributors and Fabricators Magazine
PE plastic engineer; polyethylene (UK polythene); professional engineer
PEEK polyetheretherketone
PEI polyetherimide
PEK polyetherketone
PEN polyethylene naphthalate
PES polyether sulfone
PET polyethylene terephthalate
PETG polyethylene terephthalate glycol
PEX polyethylene crosslinked pipe
PF phenol formaldehyde
PFA perfluoroalkoxy (copolymer of tetrafluoroethylene and perfluorovinylethers)
PFBA polyperfluorobutyl acrylate
phr parts per hundred of rubber
PI polyimide
PIA Plastics Institute of America
PID proportional-integral-differential
PIM powder injection molding
PLASTEC Plastics Technical Evaluation Center (US Army)
PLC programmable logic controller
PMMA Plastics Molders and Manufacturers' Association (of SME); polymethyl methacrylate (acrylic)
PMMI Packaging Machinery Manufacturers' Institute
PO polyolefin
POE polyolefin elastomer
POM polyoxymethylene or polyacetal (acetal)
PP polypropylene
PPA polyphthalamide
ppb parts per billion
PPC polypropylene chlorinated
PPE polyphenylene ether
pph parts per hundred
ppm parts per million
PPO polyphenylene oxide
PPS polyphenylene sulfide
PPSF polyphenylsulfone
PPSU polyphenylene sulphone
PS polystyrene
PSB polystyrene butadiene rubber (GR-S, SBR)
PS-F polystyrene-foam
psf pounds per square foot
PSF polysulphone
psi pounds per square inch
psia pounds per square inch, absolute
psid pounds per square inch, differential
psig pounds per square inch, gauge (above atmospheric pressure)
PSU polysulfone
PTFE polytetrafluoroethylene (or TFE)
PUR polyurethane (also PU, UP)
P-V pressure-volume (also PV)
PVA polyvinyl alcohol
PVAC polyvinyl acetate
PVB polyvinyl butyral
PVC polyvinyl chloride
PVD physical vapor deposition
PVDA polyvinylidene acetate
PVdC polyvinylidene chloride
PVDF polyvinylidene fluoride
PVF polyvinyl fluoride
PVP polyvinyl pyrrolidone

PVT pressure-volume-temperature (also P-V-T or pvT)
PW *Plastics World* magazine
QA quality assurance
QC quality control
QMC quick mold change
QPL qualified products list
QSR quality system regulation
R Reynolds number; Rockwell (hardness)
rad Quantity of ionizing radiation that results in the absorption of 100 ergs of energy per gram of irradiated material.
radome radar dome
RAPRA Rubber and Plastics Research Association
RC Rockwell C (R_c)
RFI radio frequency interference
RH relative humidity
RIM reaction injection molding
RM rotational molding
RMA Rubber Manufacturers' Association
RMS root mean square
ROI return on investment
RP rapid prototyping; reinforced plastic
RPA Rapid Prototyping Association (of SME)
rpm revolutions per minute
RRIM reinforced reaction injection molding
RT rapid tooling; room temperature
RTM resin transfer molding
RTP reinforced thermoplastic
RTS reinforced thermoset
RTV room temperature vulcanization
RV recreational vehicle
Rx radiation curing
SAE Society of Automotive Engineers
SAMPE Society for the Advancement of Material and Process Engineering
SAN styrene acrylonitrile
SBR styrene-butadiene rubber
SCT soluble core technology
SDM standard deviation measurement
SES Standards Engineering Society
SF safety factor; short fiber; structural foam
s.g. specific gravity
SI International System of Units
SIC Standard Industrial Classification
SMC sheet molding compound
SMCAA Sheet Molding Compound Automotive Alliance
SME Society of Manufacturing Engineers
S-N stress-number of cycles
SN synthetic natural rubber
SNMP simple network management protocol
SPC statistical process control
SPE Society of the Plastics Engineers
SPI Society of the Plastics Industry
sPS syndiotactic polystyrene
sp. vol. specific volume
SRI Standards Research Institute (ASTM)
S-S stress-strain
STP Special Technical Publication (ASTM); standard temperature and pressure
t thickness
T temperature; time; torque (or T_t)
TAC triallylcyanurate
T/C thermocouple
TCM technical cost modeling
TD transverse direction
TDI toluene diisocyanate
TF thermoforming
TFS thermoform-fill-seal
T_g glass transition temperature
TGA thermogravimetric analysis
TGI thermogravimetric index
TIR tooling indicator runout
T-LCP thermotropic liquid crystal polymer
TMA thermomechanical analysis; Tooling and Manufacturing Association (formerly TDI); Toy Manufacturers of America
torr mm mercury (mmHg); unit of pressure equal to 1/760th of an atmosphere

TP thermoplastic
TPE thermoplastic elastomer
TPO thermoplastic olefin
TPU thermoplastic polyurethane
TPV thermoplastic vulcanizate
T_s tensile strength; thermoset
TS twin screw
TSC thermal stress cracking
TSE thermoset elastomer
TX thixotropic
TXM thixotropic metal slurry molding
UA urea, unsaturated
UD unidirectional
UF urea formaldehyde
UHMWPE ultra-high molecular weight polyethylene (also PE-UHMW)
UL Underwriters Laboratories
UP unsaturated polyester (also TS polyester)
UPVC unplasticized polyvinyl chloride
UR urethane (also PUR, PU)
URP unreinforced plastic
UV ultraviolet
UVCA ultra-violet-light-curable-cyanoacrylate

V vacuum; velocity; volt
VA value analysis
VCM vinyl chloride monomer
VLDPE very low-density polyethylene
VOC volatile organic compound
vol% percentage by volume
w width
W watt
W/D weight-to-displacement volume (boat hull)
WIT water-assist injection molding technology
WMMA Wood Machinery Manufacturers of America
WP&RT World Plastics and Rubber Technology magazine
WPC wood-plastic composite
wt% percentage by weight
WVT water vapor transmission
XL cross-linked
XLPE cross-linked polyethylene
XPS expandable polystyrene
YPE yield point elongation
Z-twist twisting fiber direction

Acknowledgments

Undertaking the development through to the completion of the *Plastics Technology Handbook* required the assistance of key individuals and groups. The indispensable guidance and professionalism of our publisher, Joel Stein, and his team at Momentum Press was critical throughout this enormous project. The coeditors, Nick R. Schott, Professor Emeritus of the University of Massachusetts Lowell Plastics Engineering Department, and Marlene G. Rosato, President of Gander International Inc., were instrumental to the data, information, and analysis coordination of the eighteen chapters of the handbook. A special thank you is graciously extended to Napoleao Neto of Alphagraphics for the organization and layout of the numerous figure and table graphics central to the core handbook theme. Finally, a great debt is owed to the extensive technology resources of the Plastics Institute of America at the University of Massachusetts Lowell and its Executive Director, Professor Aldo M. Crugnola.

Dr. Donald V. Rosato, Coeditor and President, PlastiSource, Inc.

Preface

This book, as a two-volume set, offers a simplified, practical, and innovative approach to understanding the design and manufacture of products in the world of plastics. Its unique review will expand and enhance your knowledge of plastic technology by defining and focusing on past, current, and future technical trends. Plastics behavior is presented to enhance one's capability when fabricating products to meet performance requirements, reduce costs, and generally be profitable. Important aspects are also presented to help the reader gain understanding of the advantages of different materials and product shapes. The information provided is concise and comprehensive.

Prepared with the plastics technologist in mind, this book will be useful to many others. The practical and scientific information contained in this book is of value to both the novice, including trainees and students, and the most experienced fabricators, designers, and engineering personnel wishing to extend their knowledge and capability in plastics manufacturing including related parameters that influence the behavior and characteristics of plastics. The toolmaker (who makes molds, dies, etc.), fabricator, designer, plant manager, material supplier, equipment supplier, testing and quality control personnel, cost estimator, accountant, sales and marketing personnel, new venture type, buyer, vendor, educator/trainer, workshop leader, librarian, industry information provider, lawyer, and consultant can all benefit from this book. The intent is to provide a review of the many aspects of plastics that range from the elementary to the practical to the advanced and more theoretical approaches. People with different interests can focus on and interrelate across subjects in order to expand their knowledge within the world of plastics.

Over 20000 subjects covering useful pertinent information are reviewed in different chapters contained in the two volumes of this book, as summarized in the expanded table of contents and index. Subjects include reviews on materials, processes, product designs, and so on. From a pragmatic standpoint, any theoretical aspect that is presented has been prepared so that the practical person will understand it and put it to use. The theorist in turn will gain an insight into the practical limitations that exist in plastics as they exist in other materials such as steel, wood, and so on. There is no material that is "perfect." The two volumes of this book together contain 1800-plus figures and 1400-plus tables providing extensive details to supplement the different subjects.

In working with any material (plastics, metal, wood, etc.), it is important to know its behavior in order to maximize product performance relative to cost and efficiency. Examples of different plastic materials and associated products are reviewed with their behavior patterns. Applications span toys, medical devices, cars, boats, underwater devices, containers, springs, pipes, buildings, aircraft, and spacecraft. The reader's product to be designed or fabricated, or both, can be related directly or indirectly to products reviewed in this book. Important are behaviors associated with and interrelated with the many different plastics materials (thermoplastics [TPs], thermosets [TSs], elastomers, reinforced plastics) and the many fabricating processes (extrusion, injection molding, blow molding, forming, foaming, reaction injection molding, and rotational molding). They are presented so that the technical or nontechnical reader can readily understand the interrelationships of materials to processes.

This book has been prepared with the awareness that its usefulness will depend on its simplicity and its ability to provide essential information. An endless amount of data exists worldwide for the many plastic materials, which total about 35000 different types. Unfortunately, as with other materials, a single plastic material that will meet all performance requirements does not exist. However, more so than with any other materials, there is a plastic that can be used to meet practically any product requirement. Examples are provided of different plastic products relative to critical factors ranging from meeting performance requirements in different environments to reducing costs and targeting for zero defects. These reviews span products that are small to large and of shapes that are simple to complex. The data included provide examples that span what is commercially available. For instance, static physical properties (tensile, flexural, etc.), dynamic physical properties (creep, fatigue, impact, etc.), chemical properties, and so on, can range from near zero to extremely high values, with some having the highest of any material. These plastics can be applied in different environments ranging from below and on the earth's surface to outer space.

Pitfalls to be avoided are reviewed in this book. When qualified people recognize the potential problems, these problems can be designed around or eliminated so that they do not affect the product's performance. In this way, costly pitfalls that result in poor product performance or failure can be reduced or eliminated. Potential problems or failures are reviewed, with solutions also presented. This failure-and-solution review will enhance the intuitive skills of people new to plastics as well as those who are already working in plastics. Plastic materials have been produced worldwide over many years for use in the design and fabrication of all kinds of plastic products. To profitably and successfully meet high-quality, consistency, and long-life standards, all that is needed is to understand the behavior of plastics and to apply these behaviors properly.

Patents or trademarks may cover certain of the materials, products, or processes presented. They are discussed for information purposes only and no authorization to use these patents or trademarks is given or implied. Likewise, the use of general descriptive names, proprietary names, trade names, commercial designations, and so on does not in any way imply that they may be used freely. While the information presented represents useful information that can be studied or analyzed and is believed to be true and accurate, neither the authors, contributors, reviewers, nor the publisher can accept any legal responsibility for any errors, omissions, inaccuracies, or other factors. Information is provided without warranty of any kind. No representation as to accuracy, usability, or results should be inferred.

Preparation for this book drew on information from participating industry personnel, global industry and trade associations, and the authors' worldwide personal, industrial, and teaching experiences.

DON & MARLENE ROSATO AND NICK SCHOTT, 2011

About the Authors

Dr. Donald V. Rosato, president of PlastiSource Inc., a prototype manufacturing, technology development, and marketing advisory firm in Massachusetts, United States, is internationally recognized as a leader in plastics technology, business, and marketing. He has extensive technical, marketing, and plastics industry business experience ranging from laboratory testing to production to marketing, having worked for Northrop Grumman, Owens-Illinois, DuPont/Conoco, Hoechst Celanese/Ticona, and Borg Warner/G.E. Plastics. He has developed numerous polymer-related patents and is a participating member of many trade and industry groups. Relying on his unrivaled knowledge of the industry and high-level international contacts, Dr. Rosato is also uniquely positioned to provide an expert, inside view of a range of advanced plastics materials, processes, and applications through a series of seminars and webinars. Among his many accolades, Dr. Rosato has been named Engineer of the Year by the Society of Plastics Engineers. Dr. Rosato has written extensively, authoring or editing numerous papers, including articles published in the *Encyclopedia of Polymer Science and Engineering*, and major books, including the *Concise Encyclopedia of Plastics*, *Injection Molding Handbook 3rd ed.*, *Plastic Product Material and Process Selection Handbook*, *Designing with Plastics and Advanced Composites*, and *Plastics Institute of America Plastics Engineering, Manufacturing, and Data Handbook*. Dr. Rosato holds a BS in chemistry from Boston College, an MBA from Northeastern University, an MS in plastics engineering from the University of Massachusetts Lowell, and a PhD in business administration from the University of California, Berkeley.

Marlene G. Rosato, with stints in France, China, and South Korea, has comprehensive international plastics and elastomer business experience in technical support, plant start-up and troubleshooting, manufacturing and engineering management, and business development and strategic planning with Bayer/Polysar and DuPont. She also does extensive international technical, manufacturing, and management consulting as president of Gander International Inc. She also has an extensive writing background authoring or editing numerous papers and major books, including the *Concise Encyclopedia of Plastics*, *Injection Molding Handbook 3rd ed.*, and the *Plastics Institute of America Plastics Engineering, Manufacturing and Data Handbook*. A senior member of the Canadian Society of Chemical Engineering and the Association of Professional Engineers of Canada, Ms. Rosato is a licensed professional engineer of Ontario, Canada. She received a Bachelor of Applied Science in chemical engineering from the University of British Columbia with continuing education at McGill University in Quebec, Queens University and the University of Western Ontario, both in Ontario, and also has extensive executive management training.

Emeritus Professor Nick Schott, a long-time member of the world-renowned University of Massachusetts Lowell Plastics Engineering Department faculty, served as its department head for a quarter of a century. Additionally, he founded the Institute for Plastics Innovation, a research consortium affiliated with the university that conducts research related to plastics manufacturing, with a current emphasis on bioplastics, and served as its director from 1989 to 1994. Dr. Schott has received numerous plastics industry accolades from the SPE, SPI, PPA, PIA, as well as other global industry associations and is renowned for the depth of his plastics technology experience, particularly in processing-related areas. Moreover, he is a quite prolific and requested industry presenter, author, patent holder, and product/process developer. In addition, he has extensive and continuing academic responsibilities at the undergraduate to postdoctoral levels. Among America's internationally recognized plastics professors, Dr. Nick R. Schott most certainly heads everyone's list not only within the 2500 plus global UMASS Lowell Plastics Engineering alumni family, which he has helped grow, but also in broad global plastics and industrial circles. Professor Schott holds a BS in chemical engineering from UC Berkeley, and an MS and PhD from the University of Arizona.

CHAPTER 14
COMPRESSION MOLDING

INTRODUCTION

Compression molding (CM) encompasses different techniques in processing plastics. There is the basic CM process (Figs. 14.1 and 14.2) and the closely related transfer molding process, the compression-transfer molding process, and other molding processes. These CM methods provide different capabilities to fabricate products and to meet performance requirements using different materials (Tables 14.1 to 14.3).

CM is used to process primarily thermoset (TS) plastics. Other plastics, such as thermoplastics (TPs), elastomers, and natural rubber, are also used. CM is common; it is also the oldest method of molding TS plastics. In CM, plastic raw materials are converted into finished products by simply compressing them into the desired shapes using molds, heat, and pressure. This process can mold a wide variety of sizes and shapes, ranging from parts of an ounce to 100 lb or more.

The process requires a press with heated platens or preferably heating in the mold. Basically a two-part mold is used (chapter 17). The female, or cavity, part of the mold, when using a molding

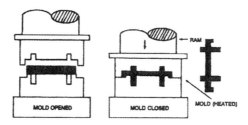

Figure 14.1 Schematic of the CM of a plastic material.

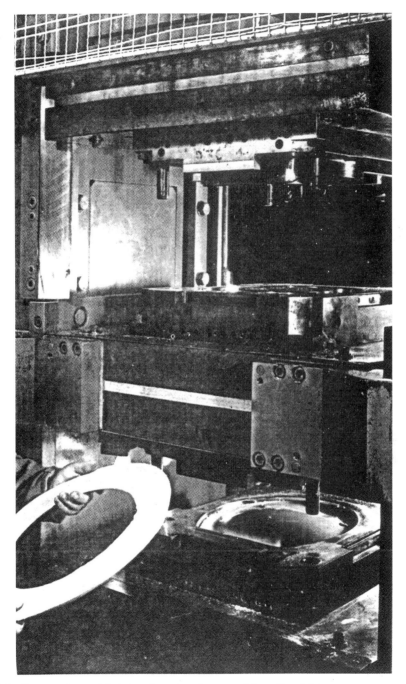

Figure 14.2 Compression molded ring-shaped part removed from the mold.

Material	Advantage	Applications
Phenol-formaldehyde	Low cost	Small housings
General-purpose	Durable	Circuit breakers
Electrical grade	High dielectric strength	Stove knobs
Heat resistant	Low heat distortion	Appliance handles, legs
Impact resistant	Strong	Kitchen appliances
Urea formaldehyde	Color stable	Plastic dinnerware
Melamine formaldehyde	Hard surface	Electrical switchgear
Alkyd	Arc resistant	Electrical switchgear
Polyester	Arc resistant	Multipin connectors
Diallyl phthalate	High dielectric strength	Encapsulating electronic components
Epoxy	Soft flowing	Encapsulating high-power electronic components
Silicone	Withstands high temperature	

Table 14.1 Example of applications for compression molded thermoset (TS) plastics

	Stamped steel	Compression molded SMC	RRIM	Injection molded thermoplastics	Stamped aluminum
Part consolidation	Baseline	Excellent	Very good	Excellent	Fair
Comparable mass	100%	75%	75%	70%	75%
Corrosion resistance	Baseline	Superior	Superior	Superior	Slight improvement
Resistance to minor impact	Baseline	Better	Best	Better	Poor
Tooling cost	100%	40%	60%	60%	100%
Raw material cost	100%	300%	600%	600%	400%
Stiffness	100%	6%	1%	2%	30%
Linear thermal	100%	100%–130%	600%–1000%	600%–1000%	170%–200%
Heat deflection temperature	N/A	Baseline	Poor	Poor	N/A

* This table used steel as a standard by which all other processes are compared

Table 14.2 Comparing compression molded properties with other processes

compound, is usually mounted on the lower platen of the press, while the male, or plunger, part is aligned to match the female part and is attached to the upper platen (Fig. 14.3). If TS-impregnated material (sheet, mat, etc.) is used, the female part of the mold is usually mounted on the upper platen of the press, while the male part is aligned to match the female part and is attached to the bottom platen (Fig. 14.4).

The plastic molding material is weighed out and is usually preheated before charging (transferring) to the cavity part of the heated mold. After charging the mold, the press is closed, bringing the two parts of the mold together. This allows the molding material to melt and flow through filling

COMPRESSION MOLDING

Process	Reinforcement wt%	Tensile Strength		Tensile Modulus		Flexural Strength		Compressive Strength		Impact Strength		Thermal Conductivity		Heat Distortion at 1.8 MPa		Dielectric Strength	
		MPa	ksi	GPa	10⁶ psi	MPa	ksi	MPa	ksi	J/m	ft·lbf/ft	W/m·K	Btu·in/h ft²·°F	°C	°F	kV/cm	kV/in.
Spray	30–50 glass-polyester	60–120	9–18	5.5–12	0.8–1.8	110–190	16–28	100–170	15–25	210–640	48–144	0.17–0.23	1.2–1.6	175–205	350–400	80–160	200–400
Compression	15–30 glass-SMC	55–140	8–20	11–17	1.6–2.5	120–210	18–30	100–210	15–30	430–1150	96–264	0.19–0.25	1.3–1.7	205–260	400–500	120–180	300–450
Compression	25–50 glass mat-polyester	170–210	25–30	6.2–14	0.9–2.0	70–280	10–40	100–210	15–30	530–1050	120–240	0.19–0.26	1.3–1.8	175–205	350–400	120–240	300–600
Filament winding	30–80 glass-epoxy	550–1700	80–250	28–62	4.0–9.0	690–1850	100–270	310–480	45–70	2150–3200	480–720	0.27–0.33	1.9–2.3	175–205	350–400	120–160	300–400
Pultrusion	40–80 glass mat-polyester	410–1050	60–150	28–41	4.0–6.0	690–1050	100–150	210–480	30–70	2400–3200	540–720	0.27–0.33	1.9–2.3	205–260	400–500	80–160	200–400
Pultrusion	30–50 glass mat-polyester	80–210	12–30	6.9–17	1.0–2.5	170–210	25–30	210–340	30–50	530–1350	120–300	0.22–0.27	1.5–1.85	95–150	200–300	80–120	200–300
Pultrusion	30–55 glass mat and roving-vinyl ester resin	70–280	10–40	6.9–21	1.0–3.0	100–280	15–40	140–340	20–50	270–1600	60–360	0.22–0.33	1.5–2.3	175–230	350–450	80–130	200–325
Pultrusion	30–55 glass mat and roving-polyester resin	50–240	7–35	5.5–17	0.8–2.5	70–210	10–30	100–280	15–40	210–1350	48–300	0.22–0.33	1.5–2.3	175–205	350–400	80–120	200–300

Table 14.3 Relating materials to properties to processes

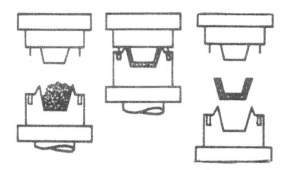

Figure 14.3 CM using a molding compound.

Figure 14.4 CM using an impregnated material.

the cavity between the two parts of the mold and at the same time pushing out any entrapped air ahead of the melt so as to fill the mold cavity completely. After holding the plastic in the mold for the time specified for a proper cure under the required temperature and pressure, the pressure is released, the mold is opened, and the solid molded plastic part is discharged. In a modern, high-speed compression press, all the operations are performed automatically.

The necessary preheating and mold-heating temperatures and mold pressure may vary considerably depending on the thermal and rheological (refers to the deformation and flow properties of the plastic) properties of the plastic (chapter 1). For a typical CM TS material, preheat may be at 200°F (93°C) and mold heat and pressure may be near 250°F to 350°F and 1000 to 2000 psi (Table 14.4). A slight excess of material is usually placed in the mold to insure that the mold is filled completely. This excess is squeezed out between the mold's mating surfaces in a thin, easily removed film known as flash. As shown in Figure 14.5, flash can form in different positions based on how it is to be removed. Different methods, such as filing, sanding, and tumbling, are used to remove flash. There are systems in which parts are frozen with dry ice, making it easier for certain types of plastic parts to be deflashed.

The molding temperature cycle for a TP starts with heating, which plasticizes the plastic, and then moves to cooling the plastic in the mold under pressure; the pressure is eventually released, and the molded article is removed. When a TS material is used, the mold does not need to be cooled at the end of the cycle, as the plastic will have "set up" and can no longer flow or distort (chapter 1).

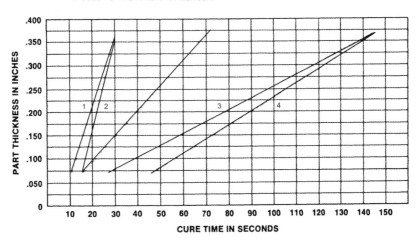

Table 14.4 Examples of the effect of preheating and part depth of phenolic parts on CM pressure (psi)

Figure 14.5 Examples of flash in a mold: (a) horizontal, (b) vertical, and (c) modified vertical

The molding cycle takes anywhere from a few minutes to an hour, depending on the type of plastic used and the size of the charge. The cycle steps are (1) charging, (2) closing the press, (3) melting the plastic, (4) applying full pressure, (5) curing for TSs or cooling for TPs, and (6) discharging or ejecting the molded part. Most of the time is consumed in the cure stage, while some of the other stages take only a few seconds.

CM was the major method of processing plastics worldwide during the first half of the last century because of the development of phenolic plastics (TSs) in 1909. By the 1940s, this situation began to change with the development and use of TPs in injection molding (IM).

About 70 wt% of all plastics were processed by CM in that early era, but by the 1950s CM's share of total production was below 25 wt%, and now that figure is about 3 wt% of all plastic products produced worldwide. Three hundred and fifty million pounds of plastic per year are consumed, so the amount going into compression molds is significant. Information on plastic consumption is in chapter 1 (Table 1.8).

This change does not mean that CM is not a viable process; it only does not provide the much lower cost-to-performance benefit of TPs that are injection molded, particularly when high production rates are called for. In the early 1900s plastics were almost entirely TS (95 wt%) used in different processes, but that proportion had fallen to about 40 wt% by the mid-1940s and now is about 10 wt%.

TSs have experienced an extremely low total growth rate, whereas TPs have expanded at an unbelievably high rate. Regardless of the present situation, CM is still important, particularly in the production of certain low-cost products, as well as in the production of heat-resistant and dimensionally precise products. CM and TM are classified as high-pressure processes, requiring 1000 to 2000 psi molding pressures. Some TSs may require higher pressures while others only require lower pressures of down to 50 psi or even just contact/zero pressure.

The advantages that keep the compression process system popular are due primarily to the simple operation that defines the system. The heated cavity is filled directly and then pressurized for the duration of the cure cycle. Examples of advantages include the following:

1. Low tooling costs because of the simplicity of the usual molds
2. Little material waste since there are usually no sprues or runners (when not compared to runnerless IM)
3. TSs that are not subject to retaining internal stresses after being cured
4. High-mechanical properties, since material receives little mastication in the process and when using reinforcements they are literally not damaged or broken
5. Less clamping pressure required compared to most other processes
6. Capital equipment that is less costly
7. Wash-action erosion of cavities that is minimal and mold maintenance that is low, since melt flow length is short

Limitations of the method include the following:

1. Fine pins, blades, and inserts in the cavity that can become damaged as the press closes when cold material is used in the cavities
2. Complex shapes that may not fill out as easily as they do in the transfer or IM processes
3. Extremely thick and heavy parts that cure more slowly than in transfer or IM (but there are preheating preforms or powder can shorten these cures)
4. TSs, with their low viscosity, that produce flash during their cure (and the flash has to be removed)

MOLD

Three types of molds are used for CM. In the positive mold (Fig. 14.6), all the material is trapped in the mold cavity. The pressure compresses the material into the smallest possible volume. Any variation in the weight of the charge will result in a variation in part thickness. In multicavity molds, if one cavity has more material than the others, it will receive proportionately greater pressure. Multiple cavities, therefore, can result in density variations between parts if they are not loaded with some degree of precision control (76, 107, 183, 210, 251, 432, 567, 568).

The mold (Fig. 14.7) is used for a variety of applications. A flash mold has a narrow land or pinchoff area around the cavity. Material is compressed in the cavity to a density that will match

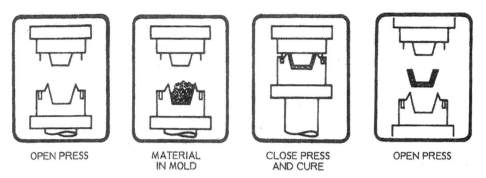

Figure 14.6 Positive compression mold.

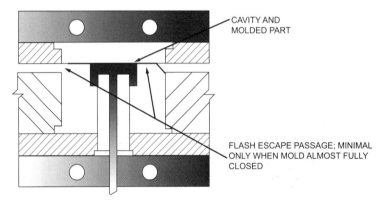

Figure 14.7 Flash compression mold.

the force applied. Excess material escapes across the pinchoff line as flash. Immediately beyond the pinchoff line, the surface is relieved to allow the flash to fluff out rather than to cure, which would result in a hard skin that would adhere tightly to the metal surface.

The semipositive mold (Fig. 14.8) is by far the most popular mold. It combines the best features of the positive molds and the flash molds. Since its design includes a plastic material well with a larger diameter and a tight-fitting force above the cavity, the material is trapped fairly positively and the plastic is forced to flow into all corners of the cavity. As the material picks up more heat and becomes fluid, it escapes between the force and cavity sidewalls as flash, allowing the force to seat on the land area.

Clearance between the cavity's sidewall and the outside diameter (OD) of the force is about 0.004 in. Variation in this clearance, which is the flash's escape channel when molding TSs, will vary the density of the molded part. The gases that are released from curing certain TS plastics and

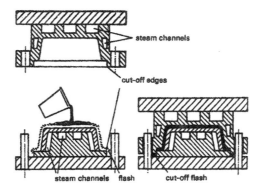

Figure 14.8 Semipositive compression mold.

the air in the cavity must be allowed to escape. They will, in some cases, filter through the flash or the clearance around the ejector pins (or both). Vents usually have to be included in the mold to permit the release of these gases. Figure 14.9 provides an example of vent locations. When TPs are processed, flash should not occur. However, air in the cavity has to be released so vents are included in the mold (Fig. 14.10).

These gases are more of a problem with urea and melamine than with phenolics. To assure they do not become entrapped in the molding material during CM and in turn weaken the molded part or cause surface blemishes, it often is advisable to open the mold to allow gases to escape. This is called a "breathing" or "bumping." It amounts to sufficient reduction in clamp pressure to allow the pressurized gases to blow their way out or sufficient opening movement to create a slight gap for trapped gases to escape effortlessly.

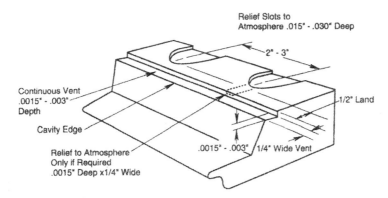

Figure 14.9 Example of mold vent locations.

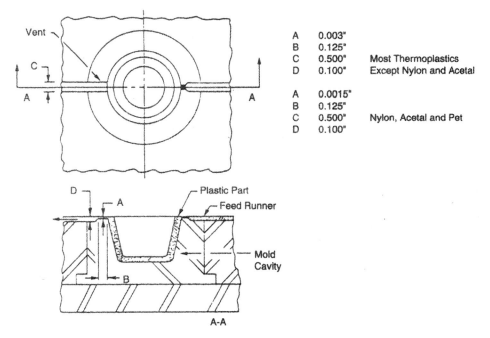

Figure 14.10 Example of vent locations in a mold processing TPs.

Lands in the mold are used to aid in controlling the thickness of molded parts or to support the pressure loads put on sections of a mold. Examples of lands are shown in Figure 14.5. Figure 14.11 shows the land locations used in a mold that supports the split-wedge in the mold.

When plastics, particularly compounds, prepregs, and sheet materials, that are filled with reinforcements, such as glass fibers, the matting edges of the mold cavity edges require special treatment. The aim is to ensure proper and clean-cut edges of the parts. The materials of construction can overlap the edges prior to or during molding. Examples of different designs used at the edges are shown in Figures 14.12 to 14.14.

MACHINES

CM machines are usually referred to as compression presses. They are primarily hydraulic or (in limited use) pneumatic. Either of these systems can have the usual straight lockup system or toggle lockup system (chapter 4). The presses may be either down-acting or up-acting. The down-acting type is used in fully automatic compression presses so that the lower half of the mold is at a fixed height to align with the feeder and stripper tables.

Various actions may be performed on molds, such as using ejector pins to remove molded parts from the cavities. Side actions may be required to remove parts that have undercuts. Other actions

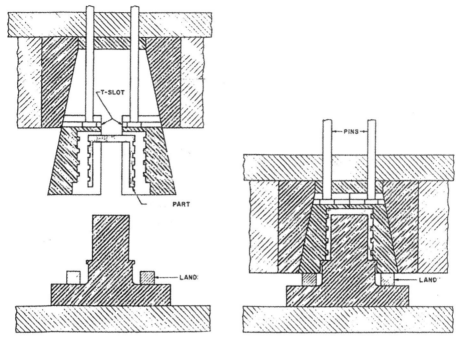

Figure 14.11 Example of land locations in a split-wedge mold (courtesy of National Tool and Manufacturing Association).

may also be required, such as as unscrewing threaded parts, including inserts. Details on mold actions are provided in chapter 17.

Presses are available in many sizes to meet different requirements. These differences may include short to long curing cycle times, small to large parts requiring different pressures (clamp tonnages), and so on. Pressures may range from less than half a ton to thousands of tons, with platens ranging from 4 × 4 in to at least 10 × 20 ft. The usual press has two platens, but some may have up to thirty platens that can simultaneously mold flat sheets or other products. There are presses with shuttle molds and others that consist of a series of individual presses (three to at least twenty-five) that rotate, allowing the TS plastic to complete its curing cycle and to ease the inclusion of inserts. Presses usually have platens parallel to each other, and there are platens that open like clamshells; these are referred to as "book type." Examples of compression presses are shown in Figures 14.15 to 14.19. Other examples are presented in chapters on the various methods (chapters 4, 12, 15, and 16).

Stamping CM presses are also in use. These presses are used for TS sheet-molding compounds (SMCs) and stampable reinforced TP sheet (STX) material. (STX, composed of a glass fiber TP, is a product of Azdel Inc., Shelby, North Carolina.) Figure 14.20 includes the sequence of compression stamping of TP sheets.

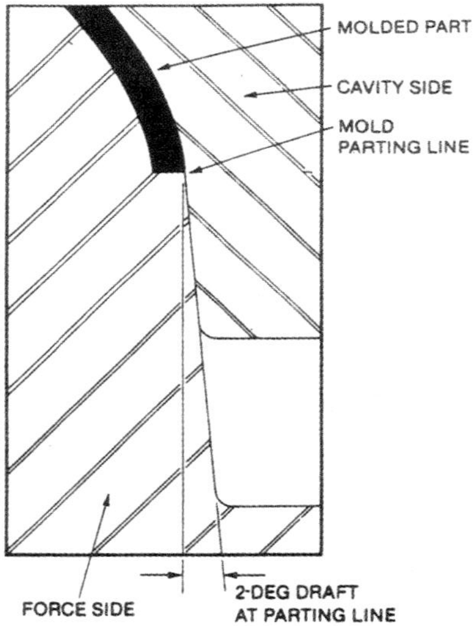

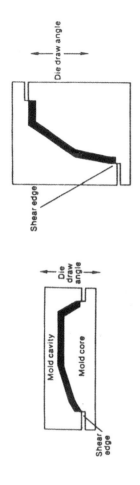

Figure 14.13 The left side is a better edge design when using a draw angle.

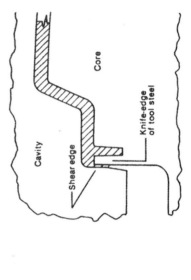

Figure 14.14 Knife shear edge.

Figure 14.12 Optimum draft for shear edges in molding sheet-molding compounds.

Figure 14.15 Press with 4 × 4 in platens and ½-ton clamp pressure (courtesy of Carver Press).

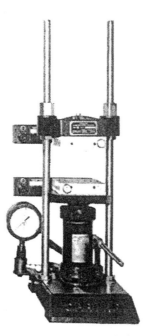

Figure 14.16 A 400-ton press with much larger than normal platens that measure 5 × 10 ft; the press has multiple zones of electrically heated platens, an automatic bump cycle, an audible alarm to signal the end of the cure cycle, and front and back safety-light curtains (courtesy of Wabash MPI).

Figure 14.17 A 4000-ton press with 5 × 8 ft platens (courtesy of Erie Press).

Figure 14.18 A 400-ton press with 18 platens, each measuring 4 × 6 ft (courtesy of Baldwin Works).

PLASTIC

Many different TS plastics are used. Phenolics, TS polyesters, diallyl phthalates, epoxies, ureas, melamines, and silicones all have their own processing requirements and resulting properties based on their compositions. (Note that there are both TS and TP polyesters.) TPs are also used (chapter 2). TSs are used primarily in CM and TPs in IM, extrusion, blow molding, and other processes. In this review, the emphasis is on TSs, which have completely different processing characteristics than TPs (chapter 1).

Figure 14.19 An 8000-ton press with 10 × 10 ft platens that have book-type opening and closing action (courtesy of Krismer, Germany).

Materials can be unreinforced resins or filler-reinforced compounds, SMCs, bulk-molding compounds (BMCs), prepregs, preforms, or mats with liquid resins, and so on. They can be TSs or TPs. TS materials cure via A-B-C stages, which identify their heat-cure cycles. As shown in Figure 14.21, in the A stage the material is uncured (as received from a material supplier), in the B stage the material is partially cured using heat, and in the C stage the material is fully cured. The typical B stage includes TS molding compounds and prepregs, which are processed to produce the fully cured

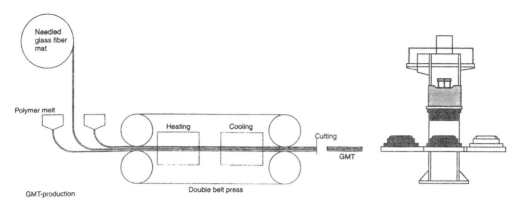

Figure 14.20 Processing sequence for compression stamping glass fiber-reinforced TP sheets.

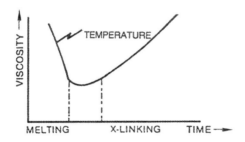

Figure 14.21 Heat-curing cycles for TPs go through A-B-C stages.

plastic material products in compression molds at the C stage. TS plastics, when heated, go through cross-linking chemical reactions to produce hard or rigid plastic products. During molding, TPs go through a melting stage when heated, followed by a hardening stage when cooled (chapter 1).

An example of a very popular CM material is bulk-molding compound; in Europe they are called dough-molding compounds (DMCs). They are formulated from different percentages of TS polyesters filled with glass fibers of lengths up to ½ in (13 mm) and fillers. The BMCs flow easily and provide high strength (chapter 15). Also popular are the TS vinyls used for phonograph records. TP vinyls are cross-linked to turn them into TS vinyls.

Very soft-flowing TS materials are required for molding around very delicate inserts. Large quantities of electronic components, such as resistors, capacitors, diodes, transistors, and integrated circuits, are encapsulated with soft-flowing TS compounds. Principally used are epoxies using CM (and transfer molding). Silicone molding compounds are used occasionally where high environmental temperatures are required of the encapsulated part, so long as the part can be exposed to temperatures of 500°F (260°C) or more. TS polyester compounds that are less expensive than epoxies and silicones are also selected when their requirements suffice (chapter 2).

In the use of preform and mat-reinforced molding, the resin may be added either before or after the reinforcement is positioned in the cavity. The preform can be a spray-up of chopped glass fibers deposited on a shaped screen with a minimum of plastic binder (about 1 to 5 wt% of resin compatible with the molding resin). Different techniques are used to provide desirable surfaces (chapter 15).

Different temperatures, pressure levels, and schedules are used, all depending on the type of material and the size and thickness of the product. Temperatures range from about 200°F to 350°F (93°C to 177°C). The range of pressure is from about 1000 to 2000 psi. Time cycles can range from less than a minute to several minutes. The process called matched-die molding generally identifies CM operating at the lower pressures.

POLYTETRAFLUOROETHYLENE BILLET

OVERVIEW

Billets are compression molded using different plastics, including those that are not the usual easy melt flowing plastics, such as polytetrafluoroethylene (PTFE). Massive (700 kg or 1500 lb), cylindrical (billet), rectangular, and sheet shapes of PTFE are produced. The blocks and cylinders can be solid or annular and are by far the heaviest objects produced from any fluoropolymer (chapter 2). The height of a cylinder can exceed 60 in (1.5 m). These billets are cut (skived) into wide, thin films (< 0.5 mm thick) or sheets (7 mm thick). Sheets, blocks, and cylinders are utilized as stock shapes for machining more complex shapes. The same principles are applied to mold any shape (146).

PTFE's specific gravity is high compared to other plastics. A solid billet measuring 300 mm tall and with a wall thickness of 130 mm may weigh about 110 pounds (50 kg). Table 14.5 presents sizes of common billets. The selection of the size of the billet depends on the properties required in the application. For example, PTFE has low thermal conductivity and thermal gradient forms across the wall thickness during sintering. Dielectric strength is more influenced than tensile strength by this thermal gradient. This is the reason thin (0.05 to 0.125 mm) electric-grade tapes are skived from billets with a wall thickness of 75 to 100 mm.

Mechanical-grade sheets are skived from heavier wall billets (125 to 175 mm). Billet height is determined by the desired width of the film or sheet. Electrical tapes are commonly made from 300 mm billets. Sheets for mechanical applications and for lining chemical processing equipment are sometimes made from 1.5 m tall billets.

The large quantity of plastic and the length of time required to produce a shape require careful attention to issues that affect productivity, such as the handling and storage of the plastic. High-temperature storage of granular PTFE can lead to compaction during handling. Plastic should be conditioned at temperatures of 70°F (21°C) to 77°F (25°C) before molding to reduce clumping

OD (mm)	ID (mm)	Wall Thickness (mm)	Weight/Height (kg/m)
500	150	175	386
500	200	150	356
400	100	150	255
480	150	150	305
180	200	140	323
300	50	125	148
300	100	100	136
250	100	75	89
200	50	75	64
150	25	62.5	37
100	20	40	16
75	35	20	7

Table 14.5 Examples of OD, ID, height, and weight relationships of different PTFE billet CMs

and ease handling. Dew-point conditions should be avoided to prevent moisture from condensing on the cold powder that will expand during sintering and crack the molding. Molding below 68°F (20°C) should be avoided because PTFE will undergo a 1% volumetric change at a transition temperature of 66.2° (19°C). Preforms molded below 68°F (20°C) can crack during sintering.

OVEN

PTFE is an excellent thermal insulator. Its thermal conductivity (0.25 W/m K), roughly 2000 times less than copper, impacts a preform's rate of sintering. The most common way of delivering heat to a preform is the circulation of hot air. A large volume of air has to be recirculated because of its low thermal capacity. Ideally, the sintering oven is electrically heated for use up to 425°C and should be equipped with override controllers to prevent overheating.

Good temperature control is critical to achieving uniform and reproducible part dimensions and properties. The interior of the oven should be designed to maximize air circulation and temperature uniformity and prevent the formation of hot spots. A highly rated insulation will minimize heat loss, which is particularly important during the sintering of a full oven load. Temperature monitoring at various locations in the oven reveals hot and cold zones, which should be corrected.

Controlled cooling is accomplished by fresh-air intake during the cool-down portion of the cycle. Very little air enters the oven during the heating part of sintering—only an amount sufficient for the removal of the off-gases. The exhaust should move directly from the oven to the atmosphere.

A hood should be placed over the oven door, where leaks are most likely to remove PTFE fumes. Adequate ventilation of the sintering area is very important. Fumes and off-gases must not be inhaled because of health hazards.

Densification and sintering

Plastic powder particles are separated by air that is removed during preforming and sintering. The powder is charged to the mold and compressed and held for a dwell period. After the preform is made, it is removed from the mold and allowed to rest for stress relaxation and degassing. The preform expands due to relaxation and recovery.

The compression pressure placed on the molded plastic exerts three types of changes in the particles of plastics. Particles undergo plastic deformation and are intermeshed together, leading to the development of cohesive or green strength. Particles also deform elastically and experience cold flow under pressure. The air trapped in the space between the particles is compressed. Removal of pressure allows the recovery of elastic deformation, which creates a quick snap-back of the preform. Over time, stress relaxation partly reverses the cold flow, and the preform expands.

The trapped air in the preform is under high pressure, theoretically equal to the preforming pressure. The air requires time to leave the preform because it is mostly contained in the void areas surrounded by enmeshed particles. Immediate sintering would lead to a rise in the air's already high pressure and catastrophic cracking of the part as the PTFE melts and the mechanical strength declines. The preform should be allowed to degas, which equalizes the internal air pressure to atmospheric pressure.

Sintering of the preform takes place in an oven, where massive volumes of heated air are circulated. Initial heating of the preform leads to thermal expansion (Fig. 14.22) of the part. After PTFE melts, relaxation of the residual stresses occurs (stored because of the application of pressure to the plastic), where additional recovery takes place and the part grows. The remaining air begins to diffuse out of the preform after heating starts. The adjacent molten particles begin to coalesce slowly; usually hours are required because of the massive size of PTFE molecules (molecular weight is 10^6 to 10^7). Fusion of the particles is followed by elimination of the voids, where almost no air is left. It is noteworthy to remember that the elimination of all the voids in PTFE is quite difficult because of limited mobility of its large polymer molecules.

Billet molding

There are three steps for producing a billet from granular PTFE: preforming, sintering and cooling.

Preforming

Preforming consists of charging the mold with the powder and compaction by the application of pressure to prepare a green part with sufficient strength to allow handling. Demolding, or removal

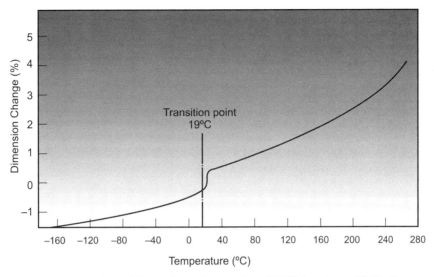

Figure 14.22 Transition point and linear thermal expansion of PTFE (courtesy of DuPont).

from the mold, and placement in the oven are the steps that require green strength. Occasionally, a preform may be machined, which increases the importance of green strength.

A key variable is resin temperature during molding. The powder is harder and has better flow below the transition temperature of 19°C (Fig. 14.22), but it does not respond well to pressure. The preforms produced below the transition temperature have low green strength and are more likely to crack during sintering. To avoid these problems, the resin should be conditioned at 21°C to 25°C for 24 hours. The molding area should ideally be maintained at > 21°C.

Adequate and uniform application of pressure is the determining factor of properties of the final part in the molding step. PTFE becomes softer and exhibits higher plastic flow as the temperature increases and can thus be molded at lower pressures. An increase of temperature from 21°C to > 31°C is roughly equivalent to 2 MPa of molding pressure. The effect of temperature is helpful, to a moderate extent, when press capacity is limited. In hot weather, a decrease in preforming pressure may be necessary to eliminate cracking problems. Economics of raised temperature molding, such as the costs involved in lengthy heat-up time to condition the resin versus the cost of additional press capacity, should be calculated before a decision is made.

Filling the mold must be done uniformly because uneven filling leads to nonuniform density in the preform and cracking. Charging the mold is much simpler with a free-flow resin than a finely cut powder. Free-flow resins more or less assume the shape of the mold and require little distribution. Conversely, significant effort has to be expended to distribute the fine-cut resin evenly in the mold. A scoop or a mesh screen should gently break up all lumps. A key consideration is to completely fill the mold prior to any pressure application.

SINTERING

A preform has limited cohesive strength and is essentially useless; sintering allows the coalescence of the resin particles, which provides strength and void reduction. Sintering cycle profiles of time and temperature affect the final properties of the billet. Sintering temperatures exceed the melting point of PTFE (342°C) and range from 360°C to 380°C.

Figure 14.23 shows the various steps of the sintering process. First, the preform completes its elastic recovery and begins to thermally expand past the PTFE melting point 342°C. The expansion can reach up to 25% to 30% by volume depending on the type of resin, powder, preforming pressure, and temperature. Above 342°C, PTFE is a transparent gel due to the absence of a crystalline phase. At the sintering temperature, adjacent melted PTFE particles fuse and coalesce. After two particles have completely coalesced, they would be indistinguishable from a larger particle, and voids are eliminated under the driving force of surface tension. Smaller particle resins and higher form pressures improve coalescence.

Coalescence and void elimination require time because of the limited mobility of PTFE molecules. Melt-creep viscosity of PTFE is in the range of 10^{11} to 10^{12} poise at 380°C, which severely inhibits any flow similar to that known for TPs. The sintering temperature is held for a period of time to allow fusion, coalescence, and void elimination to proceed and maximize properties in the part. A time is reached beyond which the part properties no longer improve and degradation begins. Property development should be balanced against cost in selecting a sintering cycle. Specific gravity increases

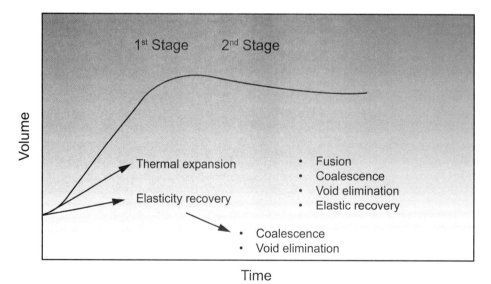

Figure 14.23 Mechanism of sintering PTFE (courtesy of DuPont).

while tensile strength decreases. Degradation of PTFE above 360°C leads to a lowering of molecular weight material, which crystallizes more easily and has decreased tensile strength.

The PTFE preform should be heated slowly because of its low thermal conductivity. This means that large volumes of turbulently heated air must be recycled through the oven to heat up the part. A thermal gradient develops between the exterior part of the preform and its interior. This gradient is required for heating the interior of the preform. Heating helps relax residual preforming stresses, which increase as the maximum pressure and mold-closing rate increase. Ideally, the slowest possible rate is best because the thermal gradient also induces mechanical stresses in the billet that, along with the residual molding stress, can surpass the cohesive strength of the preform and lead to its cracking. The economics of sintering favor the fastest rate. The compromise value is the highest heating rate, which allows relaxation of stresses in the part yet does not result in the cracking of the billet. It depends on oven temperature, airflow, and billet-wall thickness. The maximum heating rate should be determined through experiments. Table 14.6 provides recommended heating rates for preforms of various sizes.

Preform size		Sintering Cycle		
Size (mm) (dia × length) (O.D./I.D.) × (L)	Weight (kgf)	Heating rate	Sintering	Cooling rate
50 × 50	0.2	50°C/h	5 h at 370°C	50°C/h
100 × 100	1.7	30°C/h	10 h at 370°C	30°C/h
174/52 × 130	6.0	30°C/h	12 h at 370°C	30°C/h
420/150 × 600	150	50°C/h (25°C → 150°C) 3 h at 150°C 25°C/h (150°C → 250°C) 3 h at 250°C 15°C/h (250°C → 315°C) 5 h at 315°C 10°C/h (315°C → 365°C)	20 h at 365°C	10°C/h (365°C → 315°C) 10 h at 315°C 10°C/h (315°C → 250°C) 25°C/h (250°C → 100°C)
420/150 × 1200	300	50°C/h (25°C → 150°C) 5 h at 150°C 25°C/h (150°C → 250°C) 5 h at 250°C 15°C/h (250°C → 315°C) 5 h at 315°C 10°C/h (315°C → 365°C)	30 h at 365°C	10°C/h (365°C → 315°C) 10 h at 315°C 10°C/h (315°C → 250°C) 25°C/h (250°C → 100°C)

Notes: * Preforming pressure: 150 kg/cm² (dual press)
      * Compression speed: 40 to 60 mm/min (pressure applied in 4 stages)
      * Dwell time: 30 min or more

** Preforming pressure: 150 kg/cm² (dual press)
** Compression speed: 40 to 60 mm/min (pressure applied in 4 or 5 stages)
** Dwell time: 45 min or more

Table 14.6 Examples of PTFE sintering conditions

COOLING

Figure 14.24 highlights a general cooling cycle, which immediately begins at the end of the sintering time. It plays two important roles: the crystallization and the annealing of the sintered billet. Many of the properties of PTFE (similar to other semicrystalline polymers) are governed by the crystalline phase content of the part.

Crystallinity is determined by the cooling rate. At 320°C to 325°C, the molten resin reaches the freeze point and crystallization begins to take place. Polymer chains, which were randomly distributed in the molten state, begin to pack in an orderly manner during the crystallization process. The slower the cooling is, the higher the number of crystalline structures will be. This means that controlling the cooling rate can control the properties of the part. Table 14.7 shows the crystallinity

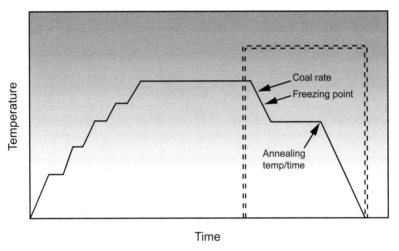

Figure 14.24 Example of a sintering cycle.

Cooling rate °C/min	% Crystallinity
Quenched in ice water	45
5	54
1	56
0.5	58
0.1	62

Table 14.7 Effect of cooling rate on crystallinity, typical for granular molding powders (courtesy of DuPont)

of the granular parts as a function of cooling rate. It is interesting that the minimum attainable crystallinity through quenching in ice water is 45% because of the low thermal conductivity of PTFE.

The actual cooling, similar to heating, is a strong function of the thermal conductivity of PTFE. Slow cooling, especially for thick parts, is necessary to avoid large thermal gradients, which can cause cracking of the part. This is especially important during the freezing transition because of the large decrease in volume that the polymer experiences while going from the melt phase to the solid phase. Large stresses are generated in the part, which can fracture the melt if the cooling rate is not sufficiently slow. The cooling rate depends on the melt strength and the wall thickness. The melt strength of a polymer increases with increasing molecular weight, and it can withstand a higher thermal gradient.

Generally, large billets (150 kg and 300 kg in Table 14.6) should be cooled at rates between 8°C and 15°C/h down to 250°C. This slow cooling rate allows the middle of the part's wall to reach the freezing point before faster cooling commences. Between 250°C and 100°C, the cooling rate can be increased to 25°C/h, and the oven doors can be opened below 100°C. Smaller parts can be cooled at higher rates below 300°C.

Annealing refers to the removal of residual stresses in the billet by holding it for a period of time between 290°C and 325°C during the cooling cycle. It also minimizes thermal gradients in the billet by allowing the wall interior to catch up with the exterior surface. The crystallinity of the part depends on the annealing temperature. A part that is annealed below the crystallization temperature range (< 300°C) will only undergo stress relief. Annealing at a temperature in the crystallization range (300°C to 325°C) results in higher crystallinity (higher specific gravity and opacity) in addition to stress relief. The part with the highest crystallinity, as evidenced by specific gravity and opacity, is obtained by sintering and annealing the part at the highest temperature.

Hot Compression-Molding PTFE

PTFE can be compression molded. Pressure and heat can be applied simultaneously followed by cooling. Sintering and cooling have to be performed in the mold. This process can be used to produce parts from unfilled and filled resins that are almost free of porosity and with unusual properties. It has been reported that increased resistance to cold flow (creep) and impact can be obtained by hot CM. The addition of a perfluorinated paraffin ($C_{25}F_{52}$) can further improve these properties. The fluorinated paraffin is to be mixed with the resin powder before charging the mold. The product of this is reported to have little tendency to flow under pressure (Table 14.8).

The mold is placed inside the oven after the preform has been made. Pressure is then applied while sintering and cooling the part. It is significantly more expensive to equip an oven for hot CM. Additional heat is also required because the melting point of PTFE increases with pressure. Filled PTFE parts are more likely to be made by this process.

Process Type	Deformation Under load, % (ASTM D621)	Notched Izod, fLlb/in (ASTM D256)	Elongation at Break, % (ASTM D638)	Tensile Strength, psi (ASTM D638)
Hot Compression Molding without $C_{25}F_{52}$	0.98	3.1	349	11,200
Hot Compression Molding with $C_{25}F_{52}$	0.87	8.3	169	2810
Standard Compression Molding	2.35	3	480	13,800

Table 14.8 Effect of CM processes on properties (courtesy of DuPont)

PROCESSING

Processing conditions such as temperature, pressure, and molding cycle differ for the different TS plastics. These molding compounds are mixtures of constituents, usually of different sizes and shapes. The compounds themselves present the greatest number of variables that must be understood and properly applied. The processing conditions with TSs and TPs ultimately affect mechanical, chemical, electrical, aesthetic, and other properties.

Many TS compounds are heated to about 300°F to 400°F (149°C to 204°C) for optimum cure, but temperatures can go as high as 1200°F (650°C). Overheating any material could degrade their performance or cause them to solidify rapidly before the cavity is completely filled.

Preheating is often used to shorten the molding cycle. It can aid in providing even heat through the material and can cause a more rapid rise in heat than occurs in the mold cavity. A warm surface plate, infrared lamps, hot-air oven, or screw/barrel preheater can accomplish preheating. The best and quickest method is high-frequency (dielectric) heating.

Preheating the compound shortens the cycle time. This is usually carried out at 150°F to 300°F (66°C to 149°C), followed by a quick transfer to a mold cavity. The actual heat depends on the material, the heater capability, and the speed of transfer. Circular preforms are used with dielectric heaters so they can be rotated to obtain uniform heating. Pills of compressed compound are used to produce preforms to reduce the bulk factor, facilitate handling, and control the uniformity of charges for mold loading. Preforms can be shaped like the mold cavity.

Compared to other processes, particularly IM for shaping plastics, CM is fairly labor-intensive even if the process is automated. However, it requires lower capital investment. Molding cycles for CM are generally longer than for IM. If the material used is preheated or preplasticized before it is placed in a mold cavity, molding cycles may be comparable to those for IM. CM flash formation may occur in the molds since the viscosity during the melting action resembles that of water. Flash can be reduced by modifying the mold design.

To aid in shortening cycle time, there are molded parts that can finish their cycles in a fixture. After a molded product is removed from the cavity, it is still hot and the material is not fully rigid.

Any internal stresses in the material may therefore cause the product's shape to change during cooling. Where close tolerances are required, and especially where products have thin sections, dimensional accuracy can be met by placing the hot, molded product on a fixture near the press, which will hold it until it has cooled (Table 14.9).

To improve mechanical properties, thermal properties, and the dimensions of certain molded TSs and TPs, exposure to a postcure can help. The part is baked in an oven. Baking times and temperature profiles required for the process can be determined through past experience or a material supplier's recommendations. Baking also improves creep resistance and reduces the number of stresses. This postcuring is also used to improve the performance of certain TPs after IM or extrusion.

Postcuring heat is usually below the molding heat. It is usually performed in a multistage heat cycle. The reinforcement system of the compound will dictate heating cycles. Products molded from compounds using organic reinforcements are postcured at lower heats than those using glass and mineral reinforcements (Table 14.10). Products of uneven thickness will exhibit uneven shrinkage. This shrinkage effect is included in the mold design.

HEATING

There are many choices in heat and temperature controls available, just as there is with other fabricating processes. They range from simple mechanical thermostats to solid-state units with proportional-integral-derivative (PID) control to microprocessors that are proportional, programmable, and self-tuning (chapter 3).

Depth of molding (in.)	Conventional phenolic		Low-pressure phenolic	
	Dielectric preheat	Not preheated	Dielectric preheat	Not preheated
$0-\frac{3}{4}$	1,000–2,000	3,000	350	1,000
$\frac{1}{4}-1\frac{1}{2}$	1,250–2,500	3,700	450	1,250
2	1,500–3,000	4,400	550	1,500
3	1,750–3,500	5,100	650	1,750
4	2,000–4,000	5,800	750	2,000
5	2,250–4,500	a	850	b
6	2,500–5,000	a	950	b
7	2,750–5,500	a	1050	b
8	3,000–6,000	a	1150	b
9	3,250–6,500	a	1250	b
10	3,500–7,000	a	1350	b
12	4,000–8,000	a	1450	b
14	4,500–9,000	a	1550	b
16	5,000–10,000	a	1650	b

Preheat(ed) units = psi

Table 14.9 Guide to wall-thickness tolerance for CM different plastics

Resin type	Production form	Cellulose paper	Cotton fabric	Asbestos paper	Asbestos fabric	Nylon fabric	Glass paper and mat	Glass fabric	Graphite	Aramid
Phenol formaldehyde	Sheet	•	•	•	•	•	•	•		
	Tube	•	•	•	•	•		•		
	Rod	•	•	•	•	•	•	•	•	
	Molded-macerated	•	•					•		
	Molded-laminated	•	•							
Melamine formaldehyde	Sheet		•		•			•		
	Tube		•		•					
	Rod		•		•			•		
	Molded-macerated									
	Molded-laminated		•				•	•		
Polyester	Sheet						•	•		
	Tube						•			
	Rod								•	•
	Molded-macerated									
	Molded-laminated						•	•	•	•
Epoxy	Sheet	•					•	•		
	Tube	•						•		
	Rod							•	•	•
	Molded-macerated									
	Molded-laminated							•	•	•
Silicone	Sheet							•		
	Tube							•		
	Rod							•		
	Molded-macerated									
	Molded-laminated									

Table 14.10 Guide in the use of reinforcements and fillers in different molding compounds

Electrical heating, through the use of heater coils, strips, or cartridges, is the most popular method. Higher temperatures for faster cycles are easily obtained. It is a cleaner system than steam, which was used many decades ago. It is important to recognize that electric mold heating is only as fast as the wattage put into it.

Steam heat was more common in the past. Steam-heat systems provide the fastest recovery time of any system because of the oversized source available in the boiler room. It offers a uniform mold temperature, as do all liquid systems, but the temperature is limited to a maximum of about 350°F. Steam heat is messy and requires good maintenance, or rusty pipes and leaks become all too common. Steam controls and the accompanying valves are expensive, and many are not dependable.

Hot-oil heat offers the benefits of higher temperature in a liquid system. It results in probably the most uniform mold temperatures, primarily because the fluid is being constantly circulated. Recovery time, however, is limited to the total heat capacity designed into the circulating unit.

High-pressure water systems that heat by continuously circulating hot water are also available. The advantage of these systems is that they produce less corrosion than steam systems because oxygen is not replenished in the closed circuit. Another advantage is that temperatures are more uniform than in steam systems because, like hot oil systems, water systems are dynamic. A disadvantage, though, is that these systems are expensive and costly to maintain.

Gas flames are used on rotary presses. Gas also has been used with some exotic materials requiring very high temperatures (over 2000°F or 1100°C).

Automation

A variety of feeders can be used to get the molding material into the mold; special strippers are also available to remove the parts. All of these tools have the common goal of faster, more efficient, and automated production.

Feeding systems include feeding cold powder through tubes from overhead hoppers; feed boards (Fig. 14.25); and feeding preheated, partially, or fully plasticated material from infrared

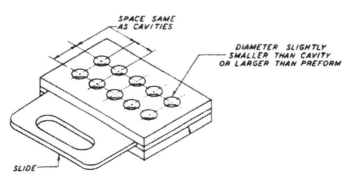

Figure 14.25 Example of a simple loading tray with a retractable slide plate to deliver material to multicavity mold.

Figure 14.26 CM machine with preplasticizer.

Figure 14.27 Three screws of the preplasticizer have been retracted from their barrels for viewing; not in the operating mode.

Figure 14.28 Preheated compounds exiting the preplasticizers prior to guillotine slicing the required shot sizes.

heated hoppers, radio frequency heating units, or screw plasticators (Figs. 14.26 to 14.28 show Stoke's machine, a machine from the past that was an excellent concept but never took off on the market). Liquid resin systems have added a further dimension, since they hasten the cure as well.

Automatic strippers have included many combinations of air blow-off, metal combs, and catch trays or chutes. Programmable robots are used for this type of work. These sophisticated units are also used to add inserts before loading the mold. The robots are extensively used to remove molded parts from the mold.

Recently all the temperature, pressure, and time controls have been replaced with a single microprocessor-based controller. There are many models available, and they allow for complex pressure and temperature curves to be programmed with multiple soaking levels, and variables can be chosen. Built-in memories recall previous programs, and cassettes can store the programs. Interfaces can connect with a central host computer for data collection or actual machine setup and supervision. The result is more flexible, more exacting, and easier to control than previous systems.

Transfer Molding

L. E. Shaw developed the plastic transfer molding process during the 1930s. It is a method of CM TS plastics, primarily. The plastic is first softened by heat and pressure in a transfer chamber (pot) and then forced by the chamber ram at high pressure through suitable sprues, runners, or gates into a closed mold to produce the molded part or parts using at least one—usually two or more—cavities (Figs. 14.29 to 14.32). Usually dielectrically preheated circular preforms are fed into the

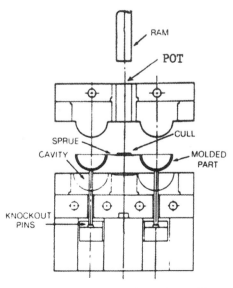

Figure 14.29 Schematic of transfer molding.

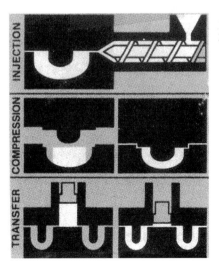

Figure 14.30 Comparing IM, CM, and transfer molding.

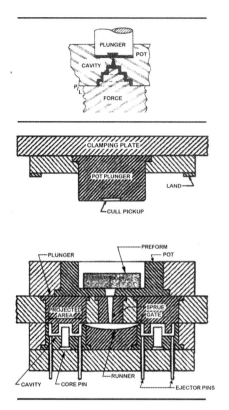

Figure 14.31 Detail view of transfer molding with two cavities.

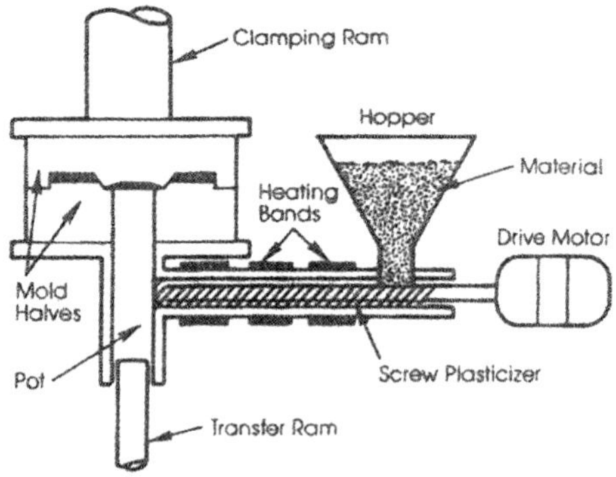

Figure 14.32 Example of a screw plasticizer preheating plastic that is delivered into the transfer molding pot for delivery into the mold cavities.

pot. Plastic remaining in the transfer chamber after the mold fills is called cull. Unless there is slight excess in this chamber, one cannot be sure that the cavity was completely full.

Since the plastic entering the cavities is melted, it requires less force to fill the cavities than CM (Tables 14.11 and 14.12). With conventional CM, there is more force in the cavities as the solid plastic is melted. The result is that more intricate parts can be molded, as can encasings for intricate devices such as electronics. Figure 14.33 shows a manual operation that is also automated in setting up electronic circuits in the multiple-cavity transfer mold.

COMPRESSION-INJECTION MOLDING

Compression-injection molding (CIM), often called injection-compression molding (ICM), is a variant of IM (chapter 4). The essential difference from IM is the manner in which the thermal contraction in the mold cavity that occurs during cooling (shrinkage) is compensated. With conventional IM, the reduction in material volume in the cavity due to thermal contraction is compensated

Characteristic	Compression	Transfer
Loading the mold	1. Powder or preforms 2. Mold open at time of loading 3. Material positioned for optimum flow	1. Mold closed at time of loading 2. Preforms RF heated and placed in plunger well
Material temperature before molding	1. Cold powder or preforms 2. Preforms RF heated to 220–280°F	Preforms RF heated to 220–280°F
Molding temperature	1. One step closures: 350–450°F 2. Others: 290–390°F.	290–360°F
Pressure via clamp	1. 2,000–10,000 psi (3,000 optimum on part) 2. Add 700 psi for each inch of part depth	1. Plunger ram at 6,000–10,000 psi 2. Clamping ram having minimum tonnage of 75% of load applied by plunger ram on mold
Pressure in cavity	Equal to clamp pressure	Very low to maximum of 1,000 psi
Breathing the mold	Frequently used to eliminate gas and reduce cure time	1. Neither practical nor necessary 2. Accomplished by proper venting
Cure time (time pressure is being applied on mold)	30–300 s but will very with mass of material, thickness of part, and preheating	45–90 s but will vary with part geometry
Size of pieces moldable	Limited only by press capacity	About 1 lb maximum
Use of inserts	Limited because inserts may be lifted out of position or deformed by closing	Unlimited but complicated; inserts readily accommodated
Tolerances on finished products	1. Fair to good: depends on mold construction and direction of molding 2. Flash = poorest, positive = best, semipositive = intermediate	Good: close tolerances are easier to hold
Shrinkage	Least	1. Greater than compression. 2. Shrinkage across line of flow is less than with line of flow

Table 14.11 Transfer molding compared to CM

by forcing in more melt during the pressure-holding phase. By contrast with CIM, a compression mold design has a male plug that fits into a female cavity, rather than the usual flat-surface parting-line mold halves employed in IM.

The melt is injected into the cavity as a short shot, thereby not filling the cavity. The melt in the cavity has no stresses; it is literally poured into the cavity. Prior to receiving the melt in the mold cavity, the mold is slightly opened so that a closed cavity exists; the male and female parts are engaged so the cavity is closed. After the melt is injected, the mold, which is on a time cycle, automatically closes, producing a relatively even melt flow. Upon controlled closing, a very uniform pressure is applied to the melt. Sufficient pressure is applied to provide a molded product without stresses. This type of molding has many advantages, one being the performances of its molded products.

Process	Limitations*
Blow molding	Limited to hollow or tubular parts; wall thickness and tolerances difficult to control; principally used with thermoplastics
Calendering	Limited to sheet materials; very thin films are not possible
Casting	Limited to simple shapes; uneconomical at high volume production rates
Centrifugal casting	Limited to simple curvatures in single-axis rotation; low production rates
Coating	Economics dependent on close tolerance control
Coining	This injection-compression process produces high pressure, stress-free precision parts
Cold-pressure molding	Limited to simple shapes and few materials
Compression molding	For intricate parts containing undercuts, side draws, small holes, delicate inserts
Encapsulation	Low volume process subject to inherent limitations on materials, which can lead to product defects
Extrusion molding	Limited to sections of uniform cross section; principally used with thermoplastics
Filament winding	Limited to shapes of positive curvature; openings and holes can reduce strength if not properly designed
Injection molding	High initial tool and die costs; not economical for small runs
Laminating	High tool and die costs; limited to simple shapes and cross section profiles
Matched-die molding	High mold and equipment costs; parts often require extensive surface finishing
Pultrusion	Close tolerance control requires care; unidirectional strength
Resin transfer molding	Low mold costs, low pressure molding, two good surfaces providing quick manufacture of wood molds and producing rather complicated small and particularly large parts to rather tight tolerances
Rotational molding	Limited to hollow parts; low production rates; principally used with thermoplastics
Slush molding	Limited to hollow parts; low production rates; limited choice of materials; principally used with thermoplastics
Thermoforming	Limited to simple parts; high scrap; limited choice of materials; principally used with thermoplastics
Transfer molding	High mold cost; high material loss in sprues and runners; and size of products limited
Wet lay-up or contact molding	Not economical for large volume production; uniformity of resin distribution difficult to control; only one good surface; limited to simple shapes

* These are general comments

Table 14.12 Transfer molding compared to reinforced plastic molding

Figure 14.33 A 64-cavity transfer mold about to receive electronic devices from a work-loading frame.

Compression and Isostatic Molding

Overview

Isostatic molding is a suitable alternative to CM techniques for the production of plastics that do not have the usual melt flow behavior, such as PTFE, which is used to make parts with complex shapes in a wide range of sizes. CM can supply a stock shape that can be machined to obtain the desired shape. The drawback to this option is the extensive machining and material costs that can drive up the total cost of the product. Isostatic molding requires relatively low-cost tooling and allows significant savings in machining and material costs. Complicated parts that require some finishing and need to be in exact or nearly exact shapes and sizes can be molded and sintered by this method. A bellows is an example of a part that can be directly molded by isostatic molding; extensive machining is required to achieve the curved contour of the bellows. Isostatic molding is the method by which all shapes of preforms can be made (146).

The isostatic technology was originally invented for ceramic and powder metal processing early in the twentieth century. It has been adopted to produce parts from granular PTFE powders. Isostatic molding (sometimes called hydrostatic molding) is another technique for producing PTFE preforms by the application of hydrostatic pressure to the powder. The powder is loaded in a closed, flexible mold. Compaction of the powder into a preform takes place when pressure is applied through the flexible part (the bladder) of the mold. The bladder is usually made of an elastomeric material, such as polyurethane. This method allows the molding of complex shapes by the placement of mandrels inside the flexible bladder.

Figure 14.34 shows the principal steps for isostatically molding a simple solid cylinder. The mold cavity is formed inside an elastomeric membrane shaped like a hollow cylinder. In this case, it does not include any mandrels and is completely filled with the powder. The elastomeric bag is closed, sealed, and placed inside a pressure vessel. The vessel containing a fluid is sealed, pressurized and held for a dwell period during which the powder is compacted by the action of a compressive pressurized fluid. At the end of the dwell time, the vessel is depressurized and the mold is removed and disassembled for the removal of the preform.

As opposed to metallic molds, the bag's flexible nature makes it difficult to define its volume. The shape of the bag may also change during the filling step unless it is supported while being charged. The change in the shape of the bag depends on several factors that include the following:

1. The original mold shape
2. Fill uniformity
3. The geometry and wall thickness of the flexible segments
4. The elastic properties of the bag
5. The fastening of the rigid and elastic sections of the mold
6. The extent and the rate of powder compaction
7. The residual stress in the bag

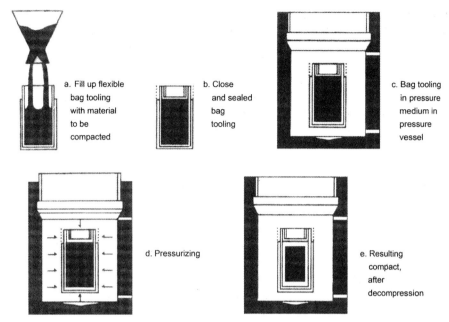

Figure 14.34 Principal steps of isostatic molding.

ISOSTATICALLY MOLDED PART	APPLICATIONS
Thin Wall Objects	Pipe liners 2.5–30 cm diameter 2.5–50 mm wall thickmess up tp 6 m length
Flanged Articles	Pump, valve and fitting liner T pieces, spacers, elbows
Closed End Articles	Cups, nose cones, radomes, bottle and caps, test tubes crucibles and laboratory ware
Encapsulation	Magnetic stirrers, butterfly valve flaps and thermocouples
Textured Objects	Beehive insulators, embossed articles, valve gate covers vessel covers and threaded parts
Solid Parts	Pyramids, balls and stopcocks,
Parts with Embedments	Reinforced mesh or sheet, bolt head or stud, conductor

Table 14.13 Examples of isostatically molded parts

Exertion of pressure on the bag is multidirectional, which conforms the resin powder to all patterns and nonuniformities in the bag. Consequently, the surfaces adjacent to the bag are less smooth than those adjacent to the smooth metallic surfaces. The importance of surface formation is one of the considerations that determine the selection of the molding process.

Isostatic molding is ideal for manufacturing thin, long objects from small tubes (5 mm diameter) or thin-walled tubes with large diameters (30 cm diameter). Examples of products made with isostatic molding include pipe liners, liners for valves and fittings, flanged parts, closed-end articles, and a host of other shapes that would require extensive machining. Table 14.13 offers examples of isostatically molded parts and their applications.

COMPARISON OF ISOSTATIC MOLDING WITH OTHER PROCESSES

Very little pressure decay occurs during isostatic compaction because of the absence of tool-wall friction. The absence of pressure decay permits the production of a great range of shapes, complexity, and sizes. This process yields a stress-free homogenous preform that exhibits uniform shrinkage during sintering. It results in a component free of distortion and with uniform physical properties. The even and constant application of pressure throughout the cycle results in a preform with lower void content. This is why shrinkage is lower and specific gravity and physical strength are higher than with techniques based on uniaxial compaction. The high extent of compaction allows the use of lower pressures in isostatic molding.

A drawback of isostatic molding is that the bag cannot produce sharp and perpendicular corners. Furthermore, the bag follows the contours of the resin particles and produces a less smooth surface than CM methods. Some machining will be required to achieve sharp corners and a smooth surface. The lower cost of isostatic molds, labor, and material renders the finished machining affordable without upsetting the economics of this process. Simple parts such as straight tubes can be made with tight tolerances, better than $\pm 2\%$.

Isostatic cycle times as short as 12 seconds are possible for small simple parts. The length of the cycle increases with the complexity and the size of the article. In such instances, this technique is often the only available method for the fabrication of those parts. An example is in-situ formation of a PTFE liner inside a T-piece fitting.

BASIC ISOSTATIC PROCESS

Solid parts can be produced by the basic process described in Figure 14.35. The only part required is a flexible bag with an end plug that can be clamped. The simplest form of the isostatic molding process entails filling a flexible molding bag with granular PTFE and inserting the filled bag in a pressure vessel containing a neutral fluid, closing the vessel, and pressurizing the fluid.

The pressure in the vessel is held for a period of time, followed by decompression, removal of the mold from the vessel, and the demolding of the preform. The preform will assume the shape of the mold and its size will depend on the extent of compaction of the PTFE powder. The part is sintered similarly to the preforms made by the other molding techniques.

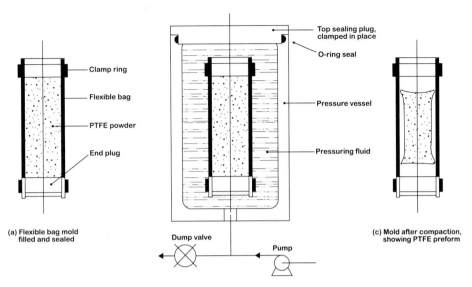

Figure 14.35 Basic isostatic compaction process.

Complex Isostatic Molding

Multipart molds are required to make more complex hollor or closed-end parts, such as tubes, liners, and beakers. Figure 14.36 depicts three methods for producing a tube-shaped PTFE preform.
These three methods are described as follows:

1. The resin is loaded between two flexible bags that, after exposure to pressure, compact from the outward and inward radial directions.
2. The outside bag is replaced with a rigid cylinder and the resin is loaded between them. The direction of compaction is radially outward, and the resin is squeezed against the surface of the outer rigid cylinder.
3. The inner bag is replaced with a rigid cylinder, and the resin is loaded between them. The direction of compaction is radially inward, and the resin is squeezed against the surface of the inner rigid cylinder.

An example of molding a different shape is a beaker. In this example, the rigid part is inside the bag. This rigid part, which is usually made of a polished metal, aids in the formation of a shape, and is often referred to as a mandrel. The surface adjacent to the mandrel assumes the surface formation of the mandrel and, in the case of a polished metal, can be very smooth. Pipe liners are normally manufactured with an outside flexible bag and an inside polished metal mandrel to produce

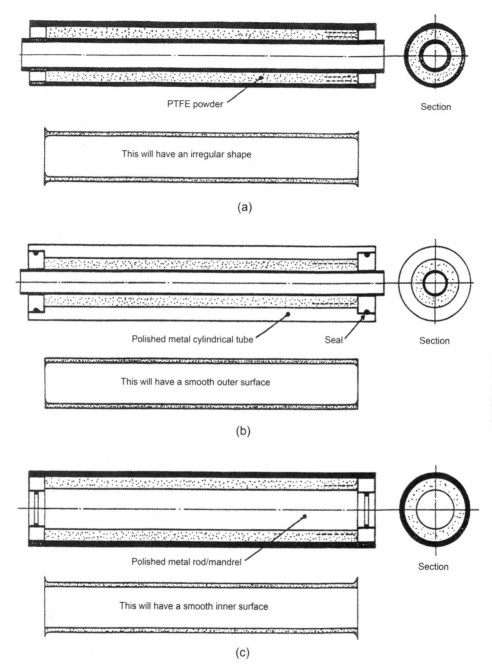

Figure 14.36 Three ways of molding PTFE tubes: (a) two flexible bags, (b) inner flexible bag with outer rigid cylinder, and (c) outer flexible bag with inner rigid rod.

a smooth surface for exposure to processing fluids. A smooth surface minimizes material trapping and builds up in the pores of the rough areas. The rigid mandrel prevents true isostatic compaction because of a friction effect, but this effect is relatively small in the majority of the molds.

Wet and dry bag molding

Wet and dry bag molding are two techniques for isostatic molding that are principally identical but operationally different. The wet bag process is similar to the basic molding procedure in which the mold is submerged in the pressurization fluid. In the dry bag technique, the mold and the bag are fixed in place and the functions of the mold and the pressurization vessel are combined. The pressurization fluid is introduced through a high-pressure liquid supply system behind the flexible bag. The mold assembly is designed to withstand this pressure. Dry bag molding contrasts with the absence of submerged mold and wet mold assembly.

The dry bag process has advantages over the wet bag process. In dry bag molding, the operation of placing the mold in the pressure vessel has been eliminated. Sealing and unsealing of the mold and the pressure vessel are reduced to just sealing and unsealing the mold. In addition to shortening cycle time, the risk of contaminating the preform with pressurization fluid is eliminated. The dry bag process can be automated and is an excellent method for large-scale production of parts.

The disadvantage of this process is the large cost differential between dry bag and wet bag molds. Dry bag molds must be able to withstand high pressure and in effect act as a pressure vessel. These molds must not be modified without reviewing the mold design because of safety considerations. It is important to take into account the need for mold modification in its initial design.

Isostatic mold design

A number of factors are fundamental to isostatic mold design regardless of the process type. Some of these factors include the preform's shape, the PTFE powder type, the surface finish of the part, the uniformity of compaction, the ease of the process, and the pressure direction.

It is important to satisfy all or as many of these considerations as possible to assure quality parts. The internal mold dimensions can be calculated from the finished size of the part, shrinkage, and compaction ratio of the resin. The rest of the design depends on the strength required to withstand wet- or dry-bag pressure. Three factors have the most important effect on the configuration and surface finish of the mold: the preform's shape, the surface finish of the part, and the pressurization direction. Table 14.14 lists some mold design considerations.

	Considerations	Some of the Factors Involved
1	Details of PTFE article to be produced	Dimensions, tolerances, surface finish, physical/chemical properties, special conditions and quantity required.
2	Approximate cost evaluation—isostatic compaction in comparison with alternative methods of production	Comparison of isostatic compaction with other techniques and quantity required.
3	PTFE powder, compaction and sintering conditions to be used	Consideration (1), and suitability for processing
4	Compaction by wet- or dry-bag technique	Estimated equipment cost (may be partly offset against future work), mold cost, cycle time, quantity required and manufacturing costs.
5	Possible directions of pressing. If both pressing from the inside and pressing from the outside are possible, consider each separately.	Direction of pressing
6	Mold configuration, i.e., basic shape and layout of components.	Good uniform compaction of PTFE Compromise between accuracy of shape and complexity of mold. Ease of mold filling and preform removal (from bag and rigid part). Bag support (mandrel or container). Good sealing. Avoidance of PTFE and bag damage. Possible problems and need for special techniques.

7	Detail design of mold rigid parts (except bag support)—dimensions, material, location of parts, seals and clamping.	(a) Molding surface dimensions:
		    Dimensions and tolerances of required article (consideration 1).
		    Allowing for PTFE shrinkage (Consideration 3)
		    Possible need for machining (Consideration 10).
		(b) Material and overall dimensions.
		    General strength, subject to size limitations for wet-bag compaction; pressure vessel design for dry-bag compaction.
		(c) Seals and clamping

8	Detail design of flexible part (bag)—dimensions, material and thickness.	(a) Dimensions:
		    Assumed bag movement, compaction ratio and shrinkage for the PTFE (Consideration 3) and rigid molding surface dimensions (Consideration 7).
		(b) Material and thickness
		    Preform surface finish, ease of preform removal, need for special techniques (e.g., pre-tensioning and cost.

9	Detail design of bag support (mandrel/container)—dimensions, bag sealing and clamping.	(a) Dimensions:
		    Bag size and thickness (Consideration 8); pressurizing fluid flow rate for transmission channel dimensions.
		(b) Bag sealing and clamping:
		    Avoidance of bag damage.

10	Processing of PTFE after sintering, e.g., machining.	Dimensional tolerances and surface finish.
11	Detail estimate of costs.	Estimated equipment cost, mold cost, cycle time, quantity, scrap rate and manufacturing cost.
12	Estimate of time delay before full-scale production.	Time for prototype or development work, and to obtain equipment, mold set.
13	Choose most suitable powder and direction of pressing if choice still remains.	Cost, quality and time delay before full-scale production.
14	Prototype	
15	Modification	Cost of modification, in relation to cost savings.
16	Final design.	

Table 14.14 Isostatic mold design considerations

Chapter 15

Reinforced Plastic

OVERVIEW

The industry continues to go through a major evolution in reinforced plastic (RP) structural and semistructural materials. RPs has been developed to produce exceptionally strong material. The RP products normally contain from 10 wt% to 40 wt% of plastic resin, although in some cases, resin content may go as high as 60% or more (Figs. 15.1 and 15.2).

RPs, also called plastic composites or composites, are tailor-made materials that provide the designer, fabricator, equipment manufacturer, and consumer engineered flexibility to meet the needs of different environments and create different shapes (Tables 15.1 and 15.2). They can sweep away the designer's frequent crippling necessity to restrict performance requirements of designs to traditional monolithic materials. The objective of a plastic composite is to combine similar or dissimilar materials in order to develop specific properties related to desired characteristics.

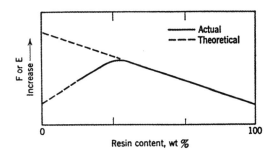

Figure 15.1 Effect of matrix content on strength (F) or elastic moduli (E) of RPs.

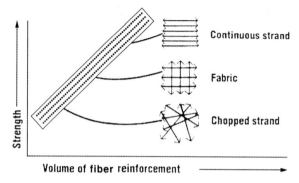

Figure 15.2 Properties versus amount of reinforcement.

Matrix material	Reinforcement material	Examples of properties modified
Thermoset plastic, Thermoplastic	Glass, aramid, carbon, graphite, whisker, etc.	Mechanical strength, wear resistance, elevated temperature resistance, energy absorption, thermal stability
Metal	Metal, ceramic, carbon, etc.	Elevated temperature strength, thermal stability, etc.
Ceramic	Metallic and ceramic particles and fibers	Elevated temperature strength, chemical resistance, thermal resistance, etc.

Table 15.1 Types of composites

Plastic-base	Ceramic-base	Elastomer-base	Metal-base
Polytetrafluoroethylene with carbon fiber	Porous oxide (silica) matrix infiltrated with phenolic resin	Silicone rubber filled with microspheres and reinforced with a plastic honeycomb	Porous refractory (tungsten) infiltrated with a low melting point metal (silver)
Epoxy-polyamide resin with a powdered oxide filler	Porous filament wound composite of oxide fibers and an inorganic adhesive, impregnated with an organic resin	Polybutadiene-acrylonitrile elastomer modified phenolic resin with a subliming powder	Hot-pressed refractory metal containing an oxide filler
Phenolic resin with an organic (nylon), inorganic (silica), or refractory (carbon) reinforcement	Hot pressed oxide, carbide, or nitride in a metal honeycomb		
Precharred epoxy impregnated with a noncharring resin			

Table 15.2 Examples of composite ablative compounds

Composites can be designed to provide practically any variety of characteristics. For this reason, almost all industries use them. Economical, efficient, and sophisticated parts are made, ranging from toys to bridges to reentry insulation shields to miniature printed circuits.

Almost any thermoset (TS) or thermoplastic (TP) matrix property can be improved or changed to meet varying requirements by using reinforcements. Typical resins used include polyester (TSs and TPs), phenolics, epoxies, silicones, diallyl phthalates, alkyds, melamine, polyamide, fluorocarbons, polycarbonate (PC), acrylics, acetal, polypropylene (PP), acrylonitrile butadiene styrene (ABS) copolymers, and polyethylene. Reinforced TSs (RTSs) predominate for the high-performance applications. However, there has been successful concentrated effort to expand use of reinforced TPs (RTPs) in electronics, automobiles, aircrafts, underground pipes (chapter 20), appliances, cameras, and other products (Figs. 15.3 to 15.11). The result is that over 50 wt% of all RPs are TP types, principally injection molded using short and long glass fibers (chapter 4).

Fiber strengths have risen to the degree that 2-D and 3-D RPs can produce very high-strength and stiff RP products having long service lives. RPs can be classified according to their behavior or performance that varies widely and depends on time, temperature, environment, and cost. The environment involves all kinds of conditions, such as amount and type of load, weather, chemical resistance, and many more. Directly influencing behaviors or performances of RPs involves the type of plastic, the type of reinforcement, and processes used. These parameters are also influenced by how the product is designed. Figures 15.12 to 15.16 and Tables 15.3 to 15.8 provide information on characteristics, properties, and processes of RPs.

DEFINITION

A precise definition of RPs is difficult or impossible to formulate because of the scale factor. At the atomic level, all elements are composites of nuclei and electrons. At the crystalline and molecular level, materials are composites of different atoms. And at successively larger scales, materials may become new types of composites, or they may appear to be homogeneous (chapter 1).

Wood is a complex composite of cellulose and lignin; most sedimentary rocks are composites of particles bonded together by natural cement; and many metallic alloys are composites of several quite different constituents. On a macro scale, these are all homogeneous materials.

In this review, RPs is considered to be combinations of materials differing in composition or form on a macro scale. But all the constituents in the plastic composite retain their identities and do not dissolve or otherwise completely merge into each other. This definition is not entirely precise, and it includes some materials often not considered to be composites. Furthermore, some combinations may be thought of as composite structures rather than composite materials. The dividing line is not sharp, and differences of opinion do exist.

The name "composite" identifies thousands of different combinations with very few that include the use of plastics. In using the term *composites* when plastics are involved, the more appropriate term is *plastic composite*.

Figure 15.3 Glass fiber-TS polyester-filament-wound RP underground gasoline storage tank.

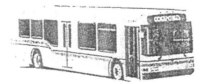

Figure 15.4 Complete primary and secondary bus structure hand layup of glass fiber-TS polyester RP.

Figure 15.5 Glass fiber swirl mat-TS polyester RP vacuum hand layup boat shell.

Figure 15.6 Glass fiber-TS polyester RP robot controlled hand layup 28 ft long boat.

Figure 15.8 Glass fiber-TS polyester filament wound RP tank trailer that transports corrosive and hazardous materials.

Figure 15.9 Pultruded glass fiber roving-TS polyester rods in a 370 ft long lift bridge supports up to 44 T traffic load.

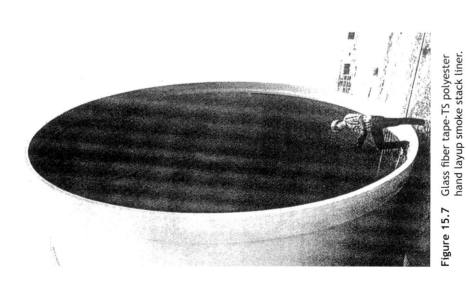

Figure 15.7 Glass fiber tape-TS polyester hand layup smoke stack liner.

Figure 15.10 Glass fiber-TS polyester filament wound RP railroad hopper car body.

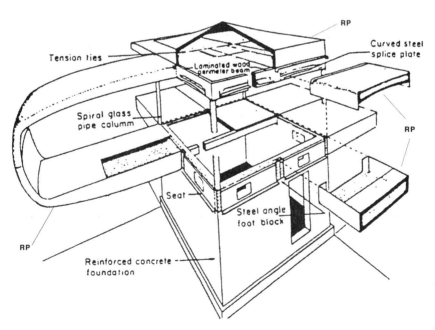

Figure 15.11 Monsanto House of the future all glass fiber-TS polyester RP hand layup has four 16 ft long U-shaped (monocoque box girders) cantilever structures 90° apart producing the main interior.

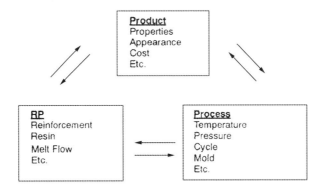

Figure 15.12 Interface of a RP.

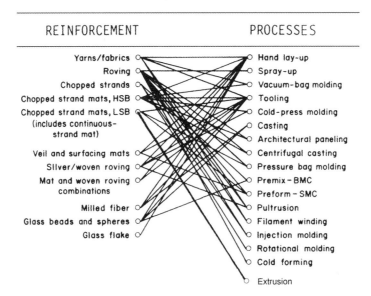

Figure 15.13 Examples of reinforcement types and processing methods.

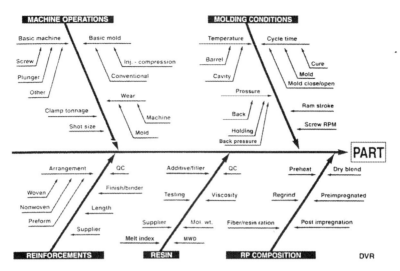

Figure 15.14 Fishbone diagram for an RP process (courtesy of Plastics FALLO).

Compression Molding	Injection Molding
Flexible Plunger	Marco Process
Flexible Bag Molding	Pultrusion
Laminate	Reactive Liquid Molding
Hand Lay-Up	Reinforced Resin Transfer Molding
Vacuum Bag Molding	Reinforced Rotational Molding
Vacuum Bag Molding and Pressure	Squeeze Molding
Pressure Bag Molding	SCRIMP Process
Autoclave Molding	Soluble Core Molding
Autoclave Press Clave	Lost-Wax Process
Wet Lay-Up	Spray-Up
Bag Molding Hinterspritzen	Stamping
Contact Molding	Cold Forming
Filament Winding	Comoform Cold Molding
Fabricating RP Tank	

Figure 15.15 Review of different processes to fabricate RP products.

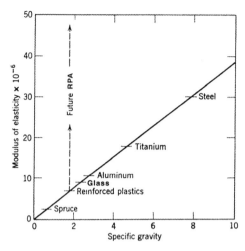

Figure 15.16 Modulus of different materials can be related to their specific gravities with RPs providing an interesting graph.

	Thermosets								Thermoplastics							
	Polyester	Polyester SMC	Polyester BMC	Epoxy	Polyurethane	Acetal	Nylon 6	Nylon 6/6	Polycarbonate	Polypropylene	Polyphenylene sulfide	ABS	Polyphenylene oxide	Polystyrene	Polyester PBT	Polyester PET
Injection molding	•		•	•	•	•	•	•	•	•	•	•	•	•	•	•
Hand lay-up	•		•													
Spray-up	•		•													
Compression molding	•	•	•	•		•				•		•				
Preform molding	•		•													
Filament winding	•		•													•
Pultrusion	•		•													
Resin transfer molding														•	•	•
Reinforced molding	•	•	•	•		•										

Table 15.3 Examples of reinforcement types and processing methods

Table 15.4 Examples of RTP properties

Plastic	Glass Fiber Content, (wt %)	Specific Gravity, D 792	Tensile Strength (MPa), D 638	Tensile Elongation (%), D 638	Tensile Modulus (GPa), D 638	Flexural Strength (MPa), D 790	Flexural Modulus (GPa), D 790	Compressive Strength (MPa), D 695	Impact Strength, Izod Notched, (J/m), D 256
ABS	10	1.10	65	3.0	4.6	102	4.5	83	64
	20	1.22	76	2.0	5.1	107	4.9	97	59
	30	1.28	90	1.4	6.3	116	6.4	104	53
Acetal	10	1.54	72	2.4	6.6	107	6.1	69	53
	30	1.63	83	2.0	7.7	114	7.2	81	43
Nylon 6	15	1.25	104	4.0	5.9	159	5.4	97	80
	30	1.37	166	3.0	7.2	200	6.9	166	117
Nylon 6/6	13	1.23	97	4.0	6.2	173	4.5	93	53
	30	1.37	173	3.0	9.0	235	9.0	186	107
Nylon 6/12	30	1.30	135	4.0	8.3	193	7.6	138	117
Polycarbonate	10	1.26	83	9.0	5.2	110	4.1	97	107
	30	1.43	121	2.0	8.6	141	6.9	117	128
Polyester, TP	30	1.52	131	4.0	8.3	193	7.9	124	96
Polyethylene	10	1.04	36	4.0	2.5	46	2.5	35	75
	30	1.18	59	3.0	5.0	89	4.9	41	91
Polyphenylene sulfide	40	1.64	152	3.0	14.1	255	13.0	145	80
Polypropylene	10	0.98	43	4.0	2.5	54	2.4	41	43
	20	1.04	45	3.0	3.7	57	3.6	45	59
	30	1.12	47	2.0	4.4	63	4.3	47	69
Polypropylene,	10	0.98	50–59	4.0	3.7	72–94	3.5	43–44	64–75
Polystyrene High heat co-polymer	20	1.22	90	1.2	8.3	131	7.9	110	59
High heat ter-polymer	30	1.35	83	1.8	6.5	123	5.7	76	80
Polysulfone	20	1.38	97	2.5	6.0	138	5.9	124	64
	40	1.55	124	1.5	11.6	173	10.7	138	80
Polyurethane	10	1.22	33	48.0	0.7	43	0.6	35	747
PVC	20	1.58	97	3.0	0.8	145	6.9	83	80
	20	1.22	100	1.8	8.6	131	7.6	121	64
SAN	35	1.35	110	1.4	10.4	155	9.3	45	53

Thermoplastic Short Forms	Average Melting Temperature	Material Structure[a]	Mold Temperature	Hot-runner Process Temp.	% Glass Fiber Weight	Mold Temperature	Hot-runner Process Temp.
PE	140	SEMI	25	250	30	40	230
PP	170	SEMI	35	255	30	40	245
PS	100	AMO	45	275	30	65	245
SAN	115	AMO	80	255	30	90	260
ABS	110	AMO	75	250	30	90	260
PMMA	100	AMO	70	245			
POM	181	CRYS	100	200	30	105	210
CA	227	AMO	75	235			
CAB	140	AMO	55	215			
CAP	190	AMO	65	225			
PET	225	CRYS	140	280	30	140	285
PC	150	AMO	90	300	40	120	310
PA 6	220	CRYS	90	250	40	110	280
PA 6/6	255	CRYS	90	285	30	110	300
PA 6/10	215	CRYS	90	250			
PA 11	175	CRYS	60	230			
PA 12	175	CRYS	60	230			
PVC	100	AMO	35	195			
PUR	160	AMO	35	205	30	50	215
PSU	200	AMO	150	315	30	160	385
PPS	290	CRYS	110	330	40	120	315
PES	230	AMO	150	350	40	150	360
FEP	275	CRYS	150	315	30	230	255
PAI	300	AMO	230	365			
PEI	215	AMO	100	370	30	150	420
PEEK	334	CRYS	160	370	30	180	380
LCP	330	CRYS	175	400	30	180	400

[a] CRYS = crystalline; AMO = amorphous; SEMI = semicrystalline. (Chapter 1).

Table 15.5 TP-glass fiber RPs injection molding (IM) temperatures

Thermosets	Properties	Processes
Polyesters	Simplest, most versatile, economical, and most widely used family of resins; good electrical properties, good chemical resistance, especially to acids	Compression molding, filament winding, hand lay-up, mat molding, pressure bag molding, continuous pultrusion, injection molding, spray-up, centrifugal casting, cold molding, encapsulation, etc.
Epoxies	**Excellent mechanical and adhesion properties, dimensional stability, chemical resistance (especially to alkalis), low water absorption, self-extinguishing (when halogenated), low shrinkage, good abrasion resistance**	Compression molding, filament winding, hand lay-up, continuous pultrusion, encapsulation, centrifugal casting
Phenolic resins	Good acid resistance, good electrical properties (except arc resistance), high heat resistance	Compression molding, continuous lamination, high pressure process
Silicones	Highest heat resistance, low water absorption, excellent dielectric properties, high arc resistance	Compression molding, injection molding, encapsulation
Melamines	Good heat resistance, high impact strength	Compression molding
Diallylphthalate	Good electrical insulation, low water absorption	Compression molding

Table 15.6 Examples of properties and processes of RTS plastics

Polyester types	Characteristics	Applications
General purpose	Rigid moldings	Trays, boats, tanks, boxes, luggage, seating
Flexible and semirigid resins	Tough, good impact resistance, high flexural strength, low flexural modulus	Vibration damping; machine covers and guards, safety helmets, electronic part encapsulation, gel coats, patching compounds, auto bodies, boats
Light-stable and weather-resistant	Resistant to weather and ultraviolet degradation	Structural panels, sky-lighting, glazing
Chemical-resistant	Highest chemical resistance of polyester group, excellent acid resistance, fair in alkalies	Corrosion-resistant pipe, tanks ducts, fume stacks
Flame-resistant	Self-extinguishing, rigid	Building panels (interior), electrical components, fuel tanks
High heat distortion	Service up to 260°C, rigid	Aircraft parts
Hot strength	Fast rate of cure (hot), moldings easily removed from die	Containers, trays, housings
Low exotherm	Void-free thick laminates, low heat generated during cure	Encapsulating electronic components, electrical parts, switchgear
Extended pot life	Void-free and uniform, long flow time in mold before gel	Large complex moldings
Air dry	Cures tack-free at room temperature	Pools, boats, tanks, containers
Thixotropic	Resists flow or drainage when applied to vertical surfaces	Boats, pools, tanks linings

Table 15.7 Properties of the popular TS polyester-glass fiber RPs

Material	Property	Glass Fiber By Weight	Specific Gravity	Density	Tensile Strength	Tensile Modulus
	Units	%		g/cm³	MPa	GPa
Glass Fiber Reinforced Thermoplastics	Polyester SMC (compression)	30.0	1.85	1.827	82.20	11.72
	Polyester SMC (compression)	20.0	1.78	1.772	36.31	11.72
	Polyester SMC (compression)	50.0	2.00	2.000	157.55	15.65
	Polyester BMC (compression)	22.0	1.82	1.799	41.10	12.07
	Polyester BMC (injection)	22.0	1.82	1.799	33.29	10.55
	Epoxy filament wound	80.0	2.08	1.910	548.00	27.58
	Polyester (pultruded)	55.0	1.69	1.661	205.50	17.24
	Polyurethane, milled fibers (RRIM)	13.0	1.07	1.052	19.18	—
	Polyurethane, flaked glass (RRIM)	23.0	1.17	1.163	30.21	—
	Polyester (spray-up/lay up)	30.0	1.37	1.356	85.61	6.895
	Polyester, woven roving (lay up)	50.0	1.64	1.633	253.45	15.51
Glass Fiber Reinforced Thermosets	Acetal	25.0	1.61	1.605	126.73	8.62
	Nylon 6	30.0	1.37	1.356	164.40	7.24
	Nylon 6/6	30.0	1.48	1.467	157.55	8.27
	Polycarbonate	10.0	1.26	1.246	82.20	5.17
	Polypropylene	20.0	1.04	1.024	44.53	3.72
	Polyphenylene sulfide	40.0	1.64	1.633	150.70	14.13
	Acrylonitrile–butadiene–styrene (ABS)	20.0	1.22	1.218	75.35	6.21
	Polyphenylene oxide (PPO)	20.0	1.21	1.190	99.33	6.34
	Polystyrene acrylonitrile (SAN)	20.0	1.22	1.218	99.33	8.62
	Polyester (PBT)	30.0	1.52	1.495	130.15	8.27
	Polyester (PET)	30.0	1.56	1.550	143.85	8.96

Table 15.8 Different properties of RTPs and RTSs per ASTM standards

Elongation	Flexural Strength	Flexural Modulus	Compressive Strength	Impact Strength IZOD	Hardness	Flammability	Specific Heat
%	MPa	GPa	MPa	kJ/m notched @ 23°C	Rockwell (except where noted)		
3.0	193.06	7.58	117.22	0.10	M79	HB	—
3.0	199.96	7.65	165.48	0.12	R121	HB	—
1.9	241.33	5.52	182.72	0.12	M95	HB	12.56
9.0	110.32	4.14	96.53	0.11	M80	V-1	12.15
3.0	57.23	3.59	172.38	0.06	R103	HB	—
3.0	255.12	13.10	144.80	0.08	R123	V-O/5V	10.47
2.0	106.87	6.00	96.53	0.06	R107	HB	—
5.0	127.56	5.17	121.35	0.10	R107	HB	8.38–16.75
1.8	131.01	7.58	120.66	0.06	R122	HB	—
4.0	193.06	8.07	124.11	0.10	R118	HB	4.61
6.6	220.64	8.62	172.38	0.10	R120	HB	—
<1.0	179.27	11.03	165.48	0.85	Barcol 68	5V	12.56
0.4	110.32	9.65	158.59	0.43	Barcol 68	5V	12.56
1.7	310.28	13.79	220.64	1.03	Barcol 68	5V	12.56
0.5	88.26	10.89	137.90	0.23	Barcol 68	5V	12.56
0.5	87.22	9.93	—	0.15	M98	VO	12.56
1.6	689.50	34.48	310.28	2.39	Barcol 50	VO	9.63
—	206.85	11.03	206.85	1.33	S.D.* 65–75	VO	11.73
140.0	—	0.26–0.37	—	—		VO	—
38.9	—	1.03	—	0.11		VO	—
1.3	193.06	5.17	151.69	0.69–0.80	Barcol 50	VO	12.98
1.6	317.17	15.51	186.17	1.75	Barcol 50	VO	—

Table 15.8 Different properties of RTPs and RTSs per ASTM standards *(continued)*

Material	Thermal Coefficient of Expansion mm/m·K × 10^{-6}	Heat Deflection (DTUL) °C @ 1820 kPa	Thermal Conductivity W/mK	Dielectric Strength mV/m	Volume Resistivity Ohm-cm	Relative Permittivity 60 Hz
Glass Fiber Reinforced Thermoplastics	8.5	161	—	22,845.	10^{14}	4.12
	2.7	211	0.84–1.64	19,695.	10^{13}	3.90
	3.2	255	0.22	15,755.	10^{15}	3.80
	3.2	141	0.66	19,695.	10^{16}	3.10
	4.3	132	1.21	17,330.	10^{15}	2.70
	2.0	266	0.29	14,970.	4×10^{15}	3.00
	3.8	99	0.20	18,315.	10^{15}	3.20
Glass Fiber Reinforced Thermosets	—	205+	—	19,695.	5.7×10^{14}	4.40
	—	205+	—	—	—	4.40
	16.9	205+	—	—	—	4.40
	11.9	260	0.70	14,770.	27×10^{14}	4.20
	11.9	260	0.70	14,770.	—	4.20
	3.6	205+	0.28	11,815.	$> 10^{12}$	—
	9.0	—	0.58	7,880.	10^{13}	4.40
	140.4	30	—	—	—	—
	95.6	—	—	—	—	—
	21.6	205+	0.22	9,850.	—	—
	7.2	205+	—	13,785.	10^{14}	4.20

Table 15.8 Different properties of RTPs and RTSs per ASTM standards *(continued)*

Arc Resistance	Water Absorption	Mold Shrinkage
seconds	Percent 24 hours	in/in
D495	D570	D955
188	.25	—
188	.10	.002
188	.50	—
190	.20	.001
190	.20	.004
—	.50	.008
80	.75	—
—	—	—
—	—	—
—	1.30	—
—	.50	—
142	.29	.004
120	1.30	.004
120	.50	.002
125	.14	.005
120	.05	.003
125	.01	.002
80	.30	.002
70	.24	.003
70	.06	.002
135	.06	.003
90	.05	.003

Table 15.8 Different properties of RTPs and RTSs per ASTM standards *(continued)*

Many combinations of reinforcements and resins are used by the industry to deliver products with diverse performance and cost characteristics. These may be in layered form, as in typical melamine-phenolic impregnated paper sheets, and polyester impregnated glass fiber mat or fabric, or in molding compound form, as in glass- or cotton-filled polyester, phenolic, or urea molding compounds. Inline compounding and injection molding (IM) TPs with long glass fibers can be performed. Glass fibers (rovings, etc.) can be fed into a single- or twin-screw extruder, where the TP is melted. It cuts the reinforcement and provides an excellent mix (383). All these resulting composites have many properties superior to the component materials.

A plastic composite is the assembly of two or more materials made to behave as a single product. Good examples of this kind of product are vinyl-coated fabric used in air mattresses, or laminated metal bonded with a plastic adhesive used in helicopter blades. The RP type of composite combines plastic resins with a reinforcing agent that can be fibrous, powdered, spherical, crystalline, or whisker, made of organic, inorganic, metallic, or ceramic material. To be structurally effective, there must be a strong adhesive bond between the resin and reinforcement.

FIBROUS COMPOSITE

The large-production reinforcing agents used today are glass, cotton, cellulosic fiber, sisal, polyamide, jute, and other synthetic fibers. Specialty agents are carbon, graphite, boron, whiskers, and steel (388). They all offer wide variations in composition, properties, fiber orientation and construction, weight, and cost (Tables 15.9 to 15.20 and Figs. 15.17 to 15.25).

Type of fiber reinforcement	Specific gravity	Density lb./in.^{3} (g/cm^3)	Tensile strength 10^3 psi (GPa)	Specific strength 10^6 in.	Tensile elastic modulus 10^6 psi (GPa)	Specific elastic modulus 10^8 in.
Glass						
E Monofilament	2.54	0.092 (2.5)	500 (3.45)	5.43	10.5 (72.4)	1.14
S Monofilament	2.48	0.090 (2.5)	665 (4.58)	7.39	12.4 (85.5)	1.38
Boron (tungsten substrate)						
4 mil or 5.6 mil	2.63	0.095 (2.6)	450 (3.10)	4.74	58 (400)	6.11
Graphite						
High strength	1.80	0.065 (1.8)	400 (2.76)	6.15	38 (262)	5.85
High modulus	1.94	0.070 (1.9)	300 (2.07)	4.29	55* (380)	7.86
Intermediate	1.74	0.063 (1.7)	360 (2.48)	5.71	27 (190)	4.29
Organic						
Aramid	1.44	0.052 (1.4)	400 (2.76)	7.69	18 (124)	3.46

Table 15.9 Properties of fiber reinforcements

Property	E glass	Carbon	HM carbon	Aramid
Fiber diameter, μ (mil)	10–17 (0.39–0.67)	7 (0.27)	8 (0.31)	12 (0.47)
Therm. Cond., BTU-in./hr.-ft.^{2} (W/m·K)	7.0 (1.0)	60 (8.6)	97 (14)	3.5 (0.50)
Specific Heat @ 70 F, BTU/lb. /°F (J/Kg·K)	0.192 (803)	0.17 (710)	0.17 (710)	0.34 (1,400)
Coefficient of thermal expansion 10^{-6} in./in./ F (10^{-6} cm/cm C)				
Longitudinal	1.6 (2.9)	−0.55 (−0.99)	−0.28 (−0.50)	−1.1 (−2.0)
Transverse	4.0 (7.2)	9.32 (16.8)	— (1.8)	33.0 (59.4)
Surface energy, ergs/cm^2	31.0	53.0	—	41.0

Table 15.10 Reinforcement thermal properties

Reinforced Plastic

Plastic	Glass-fiber Content, wt %	Specific Gravity, D 792	Tensile Strength, MPa[a] (psi) D 638	Tensile Elongation, %, D 638	Tensile Modulus, GPa[a] (kips/in²) D 638	Flexural Strength, MPa[a] (psi) D 790	Flexural Modulus, GPa[a] (kips/in²) D 790	Compressive Strength, MPa[a] (psi) D 695	Impact Strength, Izod Notched, J/m[b] D 256 (ft-lbf/in.)
ABS	30	1.28	90 (13,000)	1.4	6.3 (910)	116 (16,800)	6.4 (930)	104 (15,100)	53 (0.99)
acetal	30	1.63	83 (12,000)	2.0	7.7 (1,100)	114 (16,500)	7.2 (1,050)	81 (11,700)	43 (0.80)
nylon-6	30	1.37	166 (24,000)	3.0	7.2 (1,050)	200 (29,000)	6.9 (1,000)	166 (24,100)	117 (2.19)
nylon-6/6	30	1.37	173 (25,100)	3.0	9.0 (1,300)	235 (34,100)	9.0 (1,300)	186 (27,000)	107 (2.00)
nylon-6/12	30	1.30	135 (19,600)	4.0	8.3 (1,200)	193 (28,000)	7.6 (1,100)	138 (20,000)	117 (2.19)
polycarbonate	30	1.43	121 (17,500)	2.0	8.6 (1,250)	141 (20,400)	6.9 (1,000)	117 (17,000)	128 (2.40)
polyester, thermoplastic	30	1.52	131 (19,000)	4.0	8.3 (1,200)	193 (28,000)	7.9 (1,100)	124 (18,000)	96 (1.8)
polyester, thermoplastic	30	1.52	131 (19,000)	4.0	8.3 (1,200)	193 (28,000)	7.9 (1,100)	124 (18,000)	96 (1.8)
polyethylene	30	1.18	59 (8,600)	3.0	5.0 (720)	89 (12,900)	4.9 (710)	41 (5,900)	91 (1.7)
poly(phenylene sulfide)	40	1.64	152 (22,000)	3.0	14.1 (2,050)	255 (37,000)	13.0 (1,900)	145 (21,000)	80 (1.5)
polypropylene	30	1.12	47 (6,800)	2.0	4.4 (640)	63 (9,100)	4.3 (620)	47 (6,800)	69 (1.3)
polypropylene, chemically coupled	30	1.12	68–83 (9,900–12,000)	2.0	4.6 (670)	90–131 (13,000–19,000)	4.6 (670)	45–48 (6,500–7,000)	69–91 (1.3–1.7)
polystyrene high heat copolymer	20	1.22	90 (13,000)	1.2	8.3 (1,200)	131 (19,000)	7.9 (1,100)	110 (16,000)	59 (1.1)
high heat terpolymer	30	1.35	83 (12,000)	1.8	6.5 (940)	123 (17,800)	5.7 (830)	76 (11,000)	80 (1.5)
polysulfone	40	1.55	124 (18,000)	1.5	11.6 (1,680)	173 (25,100)	10.7 (1,550)	138 (20,000)	80 (1.5)
polyurethane	10	1.22	33 (4,800)	48.0	0.7 (100)	43 (6,200)	0.6 (90)	35 (5,100)	747 (14.0)
PVC	20	1.58	97 (14,000)	3.0	0.8 (116)	145 (21,000)	6.9 (1,000)	83 (12,000)	80 (1.5)
SAN	35	1.35	110 (16,000)	1.4	10.4 (1,510)	155 (22,500)	9.3 (1,350)	45 (6,500)	53 (0.99)

Table 15.11 Properties of glass-fiber RPs

Property	Aramid[a]	Type HT carbon	E-HTS glass
Tensile strength, 10³ p.s.i.	525	450	350
Modulus, 10⁶ p.s.i.	18	32	10
Elongation to break, %	2.5	1.40	3.5
Density, lb./in.³	0.052	0.063	0.092

a—Kevlar 49.

Table 15.12 Comparative yarn properties

Fiber types	Young's modulus range, 10⁶ p.s.i.	Tensile strength range, 10³ p.s.i.	Raw material employed
Low modulus	5-10	50-150	Rayon, pitch
Intermediate modulus	30-40	450-800	PAN
High modulus	50-60	300-400	PAN, pitch
Very high modulus	70-100	250-350	PAN, pitch

Table 15.13 Examples of different carbon fibers

Base Resin	Specific Gravity D 792	Mold Shrinkage (in./in.) D 955	Water Absorption, 24-hr. (%) D 570	Tensile Strength 10³ psi (MPa) D 638		Flexural Modulus 10⁶ psi (GPa) D 790	Impact Strength, Izod (ft.-lb./in.)		Thermal Expansion (10⁻⁵ in./in.-°F) D 696	Deflection Temperature, 264 psi °F (°C) D 648
							Notched D 256	Unnotched D 256		
Nylon 6/6	1.19 (1.14)	0.008 (0.016)	0.90 (1.50)	14.5 (11.8)	100.0 81.4	0.64 (4.4) (0.41) (2.8)	1.0 (0.9)	6.7 —	2.4 (4.5)	450 (232) (170) (76.7)
Nylon 6/6	1.29	0.008	0.6	(13.5)	93.1	0.55 (3.8)	1.0	8.5	3.1	465 (240)
Nylon 6	1.19 (1.14)	0.008 (0.016)	1.0 (1.8)	13.0 (11.8)	89.6 81.4	0.58 (4.0) (0.40) (2.8)	1.1 (1.0)	9.0 —	3.0 (4.6)	390 (199) (167) (75)
Polyester (PBT)	1.33 (1.31)	0.013 (0.020)	0.06 (0.08)	9.5 (8.5)	65 59	0.60 (4.1) (0.34) (2.3)	0.8 (1.2)	9.0 —	3.0 (5.3)	380 (193) (130) (54.4)
Polycarbonate	1.23 (1.20)	0.005 (0.006)	0.12 (0.15)	11.0 (9.0)	75.8 62	0.54 (3.7) (0.33) (2.3)	0.9 (2.7)	11 (60)	3.0 (3.7)	280 (138) (265) (129)

Table 15.14 Aramid fiber-TP RP properties

Resin type		PA	PA	PA 6	PA 66
Reinforcement/ structure		40% glass fiber	40% glass fiber	30–50% glass fiber	30–60% glass fiber
US figures are given in tinted panels					
Glass transition temperature	°C °F				
Density/ specific gravity	g/cm³ lb/in³				
Mechanical properties					
Tensile strength	MPa 10³ psi	230 33	196 28	166–248 23.7–35.4	176–283 25.2–40.5
Tensile modulus	GPa 10⁶ psi		12.9 1.7	9.8–16.7 1.3–2.2	1.4–2.9
Elongation at yield	%			2.1	2.1
Bending/ flexural strength	MPa 10³ psi	343 49	300 43	269–377 38.4–53.9	283–455 40.4–65.0
Bending flexural modulus	GPa 10⁶ psi	1.6	43	9.1–15.2 1.2–20	9.1–19.7 1.2–2.6
Compressive strength	MPa 10³ psi			206–273 29.5–39.0	218–329 31.2–47.0
Izod impact strength	ft lbs/in	6.6	4.2	4.2–8.4	4.2–100
Thermal properties					
Coefficient of linear thermal expansion	k⁻¹ x10⁻⁵ in/in/°F				
Thermal conductivity	W/mk btu(h.ft²/°F)				
Heat deflection temperature	°C x 1.8 MPa °F x 264 psi			204–213 400–415	252–263 485–505
Continuous use temperature	°C °F				
Flammability properties					
Spread of flame	BS 476				
Burn rate (ISO 3795)	mm/min				
Limiting Oxygen Ind	%				
UL flammability	1.3–10mm				
Electrical properties					
Tracking resistance	Volts				
Comp. tracking resist	CTI				
Surface resistivity	Log₁₀ Ohm Ohms/sq				
Volume resistivity	Ohms-cm				
Dissipation factor	1MHz				
Dimensional change					
Water absorption	%				
Mold shrinkage	in/in				

Table 15.15 Properties of unidirectional hybrid-nylon RPs

Ratio of aramid[a]/carbon[b]	Dynamic flexure strength, 10^3 psi	Impact energy, ft.-lb./in.^{2}
100/0	63	48
50/50	82	44
25/75	82	34
0/100	99	28

Table 15.16 Charpy impact test results of square woven fabric using hybrid fibers-nylon RPs

Nominal notch length, in.	Net failure stress, 10^3 lb./in.^{2}		Stress concentration factor, K		% initial strength retained	
	Aramid[a]	E-glass[b]	Aramid	E-glass	Aramid	E-glass
0	35.4	31.3	—	—	—	—
0.25	33.5	26.4	1.06	1.19	94.6	84.3
0.50	31.3	22.7	1.13	1.38	88.4	72.5
1.00	30.0	21.3	1.18	1.47	84.7	68.1

a—Style 1350 woven roving of Du Pont's Kevlar 49 aramid on either side of 1.5 oz./ft.^{2} glass CSM. Resin: Reichhold's 33-072 polyester.
b—24 oz./yd.^{2} glass fiber WR on either side of 1.5 oz./ft.^{2} glass CSM. Resin: Reichhold's 33-072 polyester.

Table 15.17 Damage propagation of aramid and E-glass RPs using tensile-notched test specimens

Property	Unit	Woven cloth	Chopped strand mat	Continuous roving
Glass content	%	55	30	70
Specific gravity		1.7	1.4	1.9
Tensile strength	MPa	300	100	800
Compressive strength	MPa	250	150	350
Bend strength	MPa	400	150	1000
Modulus in bend	GPa	15	7	40
Impact strength: Izod (unnotched)	kJ/m^2	150	75	250
Coefficient of linear thermal expansion	$\times 10^{-6}$ per °C	12	30	10
Thermal conductivity	W/mK	0.28	0.2	0.29

Table 15.18 Examples of different glass fiber yarns

Fiberglas Yarn Number	Final Twist (T.P.I.)	Approx. Yards/ Pound	Minimum Breaking Strength (lbs.)*	Fiberglas Yarn Number	Final Twist (T.P.I.)	Approx. Yards/ Pound	Minimum Breaking Strength (lbs.)*
ESE 70/1-R	8.5Z	7,000	1.7	ESE 31/2	6.5S	1,586	9.6
ESE 60/1-R	9.4Z	6,000	2.8	ESE 31/3	4.0S	1,040	18.4
ESE 50/1-R	8.5Z	5,000	2.8	ESE 25/2	6.5S	1,285	9.2
ESE 40/1-R	8.5Z	4,000	3.4	ESE 19/2	5.5S	975	16.0
ESE 31/1	8.5Z	3,172	4.0	ESE 12.5/2	5.5S	625	14.8
ESE 25/1	8.5Z	2,570	4.7	ESE 10/2	3.5S	488	21.0
ESE 19/1	7.0Z	1,950	6.8	ESE 8.4/2	3.5S	420	21.6
ESE 12.5/1	4.0Z	1,250	9.0	ESE 6.2/2	3.5S	313	48.0
ESE 10/1	4.0Z	975	10.0	ESE 1.5/2	2.0S	70	124.0
ESE 8.4/1	4.0Z	840	10.8	ESG 12/2	3.5S	580	16.9
ESE 6.2/1	4.0Z	624	15.0	ESG 10/2	3.5S	488	22.6
ESE 1.5/1	2.0Z	140	48.0	ESG 9/2	3.5S	435	25.4
ESG 12/1	4.0Z	1,170	9.6	ESG 8/2	3.5S	390	29.5
ESG 10/1	4.0Z	975	11.3	ESG 6/2	3.5S	285	40.5
ESG 9/1	4.0Z	870	12.7	ESG 5/2	2.0S	240	42.2
ESG 8/1	4.0Z	780	12.0	ESG 4/2	2.0S	190	56.5
ESG 6/1	4.0Z	580	16.5	ESG 2.8/2	2.0S	130	55.6
ESG 5/1	2.0Z	490	19.5	ESG 1.4/2	2.0S	62	87.8
ESG 4/1	2.0Z	390	24.0	CSE 44/1-R	8.5Z	4,380	4.3
ESG 2.8/1	2.0Z	260	27.5	CSE 25/1	8.5Z	2,570	4.7
ESG 1.4/1	2.0Z	124	51.0	CSE 12.5/1	4.0Z	1,250	9.0
ESE 70/2-R	6.5S	3,500	4.3	CSE 44/2-R	6.5S	2,160	8.6
ESE 60/2-R	7.2S	3,000	5.6	CSE 25/2	6.5S	1,285	9.2
ESE 50/2-R	6.5S	2,500	5.4	CSE 12.5/2	3.5S	625	14.8
ESE 40/2-R	6.5S	2,000	7.2				

*"Minimum Breaking Strength, Pounds" values are given only as a guide not as a specification. Values above were determined by the method outlined in A.S.T.M. D578. Type A Machines are used for testing yarns where breaking strength is 50 lbs. or over; Type B Machines are used to test yarns where breaking strength is under 50 lbs.

**"Approximate Build-up" values are average thickness increases as measured on mandrels. They will vary with tension and method of application. Figures are single-wall build-up.

Table 15.19 Examples of glass fiber staple fiber yarn data

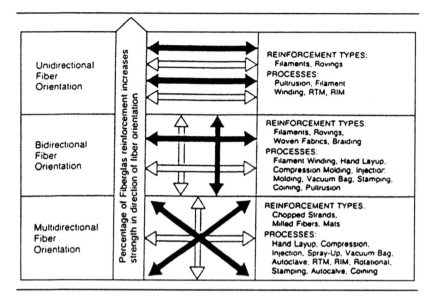

Table 15.20 Examples of glass fiber cloth constructions

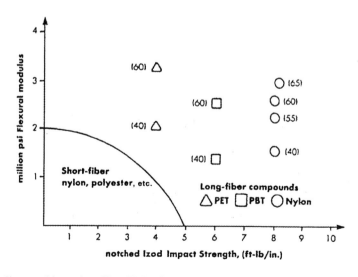

Figure 15.17 Short and long glass fiber-TP RP data (wt% fiber in parentheses).

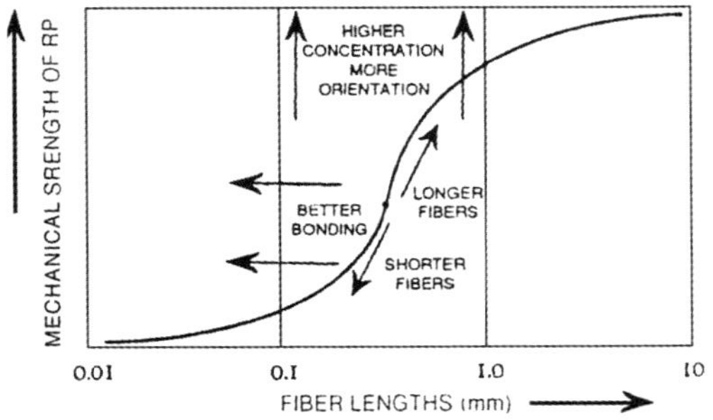

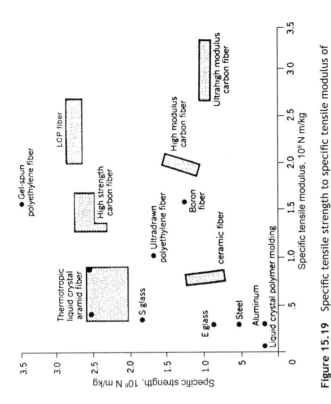

Figure 15.18 Short to long fibers influence properties of RPs.

Figure 15.19 Specific tensile strength to specific tensile modulus of elasticity data f nylon RPs.

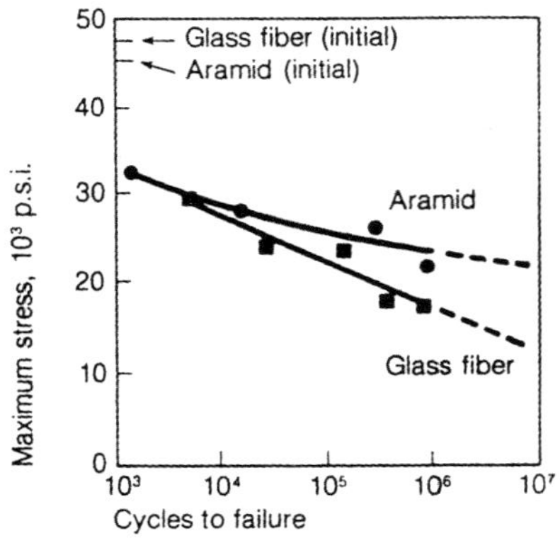

Figure 15.20 Flexural fatigue data of woven glass fiber roving RPs.

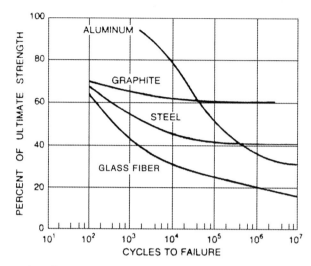

Figure 15.21 Common glass fiber-TS polyester resin RP fatigue data versus other materials (chapter 19).

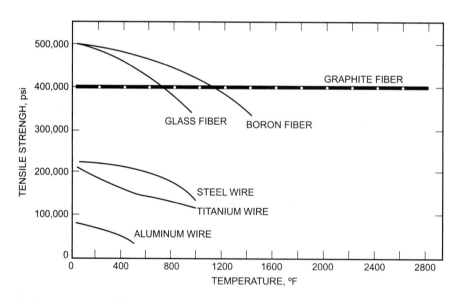

Figure 15.22 Comparing different fiber material strength properties at elevated temperatures.

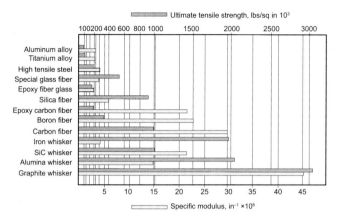

Figure 15.23 Comparing whisker reinforcements with other reinforcements.

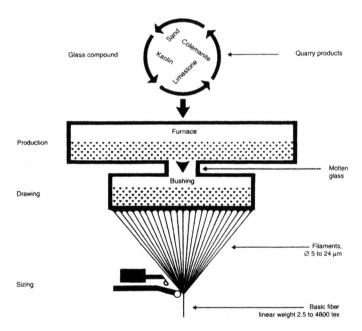

Figure 15.24 Schematic example in the manufacture of glass filaments/fibers.

Glass fibers, the most widely used at over 90 wt% of all reinforcements, are used in many forms for producing different commercial and industrial products, as well as parts in space, aircraft, surface water, and underwater vehicles. The older and still popular form is E-glass. S-glass produces higher strength properties. Other forms of glass fiber exist that meet different requirements.

Materials in the form of fibers are often vastly stronger than the same materials in bulk form. Glass fibers, for example, may develop tensile strengths of 7 MPa (1,000,000 psi) or more under laboratory conditions, and commercial fibers attain strengths of 2800 to 4.8 MPa (400,000 to 700,000 psi), whereas massive glass breaks at stresses of about 7 MPa (1000 psi). The same is true of many other materials, whether they are organic, metallic, or ceramic.

Acceptance and use of nonwoven fabrics as reinforcement for structural plastics continues to increase. Only with nonwoven fiber sheet structures can the full potential of fiber strength be realized (45). Great advances have been made in developing new fibers and resins, in new chemical finishes given to the fiber, in methods of bonding the fiber to the resin, and in mechanical processing methods. Nonwoven fabrics are inherently better able to take advantage of these developments than are woven sheets.

The strength of commercial RPs is far below any theoretical strength. Ordinary glass fibers are three times stronger and stiffer for their weight than steel. Nonwoven glass fiber structures usually have strength about 40% to 50% below that of woven fabric layups. But in special constructions, properly treated fibers have produced laminates as strong as the woven product, or in some cases even stronger.

RPs are usually applied as laminates of several layers. Many variables are important in determining the performance of the finished product. Some of the important variables are the orientation of plies of the laminate, the type of resin, the fiber-resin ratio, the type or types of fibers, and the orientation of fibers.

Nonwoven fabrics are fibrous sheets made without spinning, weaving, or knitting. They include felts, bonded fabrics, and papers. The interlocking of fibers is achieved by a combination of mechanical work, chemical action, moisture, and heat by either textile or paper-making processes.

Still stronger and stiffer forms of fibrous materials are the unidirectional crystals called whiskers (Fig. 15.23). Under favorable conditions, crystal-forming materials will crystallize as extremely fine filamentous single crystals a few microns in diameter and virtually free of the imperfections found in ordinary crystals. Whiskers are far stronger and stiffer than the same material in bulk form.

Fine filaments or fibers by themselves have limited engineering use. They need support, something to hold them in place in a structure or device. This is accomplished by embedding the fibers in a continuous supporting matrix sufficiently rigid to hold its shape, to prevent buckling and the collapse of the fibers, and to transmit stress from fiber to fiber. The matrix may be, and usually is, considerably weaker, of lower elastic modulus, and of lower density than the fibers. By itself it would not withstand high stresses. When fibers and a matrix are combined into a composite, a combination of high strength, rigidity, and toughness frequently emerges that far exceeds these properties in the individual constituents.

Polyamide (nylon) reinforcements most often are fabric and provide excellent electrical-grade laminates for conventional industrial use. It has low water absorption, good abrasion resistance, and resistance to many chemicals.

Carbon and graphite fibers are made by the pyrolysis of certain naturally occurring and man-made fibers, such as regenerated cellulose (rayon) fibers. A wide range of physical, mechanical, and chemical properties may be obtained, dependent on the amount of dehydration. This product is one of the most structurally efficient reinforcements. Unlike any other reinforcement, it retains its 2800 MPa (400,000 psi) tensile strength when tested up to a temperature of 4800°F (2700°C).

Boron in high-modulus and strength properties is available with this type of fiber. A vapor deposition process is the principal method to produce boron filaments, using ½ mil tungsten wire as a plating substrate.

Laminar Composite

Combining layers of materials into a laminated composite is an ancient art, as illustrated by Egyptian plywood, Damascus and Samurai swords, and medieval armor. There are many reasons for laminating; among them are superior strength, often combined with minimum weight; toughness; resistance to wear and corrosion; decoration; safety and protection; thermal or acoustical isolation; color and light transmission; shapes and sizes not otherwise available; controlled distortion; and many others.

Many processes involving temperature fluctuations are made to be self-regulating by employing laminates of two metals with different coefficients of expansion. When a strip of such metal changes temperature, the different expansivities of the two metals cause the strip to bend, rotate, or elongate, depending on its shape. In so doing, it can make or break electrical contacts, control the position of a damper, or perform many other functions. These bimetals or thermostat metals are servomechanisms; they respond to environmental stimuli to provide self-regulating behavior. They have this ability because they are composites; each metal by itself would not exhibit this behavior.

High-strength aluminum alloys are frequently deficient in corrosion resistance. High-purity aluminum and certain aluminum alloys are considerably more resistant to corrosion but are deficient in strength. By applying surface layers of the corrosion-resistant metal to a core of the strong alloy, a clad aluminum composite is created that has corrosion resistance that cannot be obtained using either constituent alone.

Window glass by itself is hard and durable but brittle, and upon impact may shatter into lethal shards. Polyvinyl butyral by itself is a tough but limp and easily scratched plastic material unsuitable for windows. When it is laminated between two sheets of glass, the result is a composite in which the tough plastic layer (polyvinyl butyral), firmly bonded to the glass, prevents the shards from flying when the sheet is struck. Safety glass is thus a composite laminate having properties unattainable by the constituents alone while offering the most valuable characteristics of each. This product has been produced since the early 1930s.

Some laminates are so common that they are household words. Among them are the plastic composites consisting of layers of heavy and strong Kraft paper impregnated with phenolic resins.

The resulting sheet is serviceable for many mechanical and electrical purposes. When combined with a melamine-formaldehyde–saturated overlay sheet, a familiar decorative sheet that is widely used for counters, furniture, and wall covering. Paper of three types are in common use: Kraft paper (which has high strength when compared to other papers), alpha paper (used in the electrical industry), and rag paper (which has low moisture pickup with good machinability).

For heavy-duty purposes such as bearings, tough and strong fabrics such as cotton duck are substituted for the paper. Fabric-based laminates may be further modified with graphite, fluorocarbons, or other low-friction materials to create low-friction composite bearings requiring no lubricant.

Particulate Composites

Particulate composites are used in greater volume than any others; concrete is a particulate composite. In many ways, concrete is the archetype of this class of composites. It consists of particles or aggregates of various sizes and almost always of mineral materials, bonded by a matrix of an inorganic cement originally mixed with and hardened by its chemical reaction to water. Many types of particles are employed; at least five different types of Portland cements and several other types of inorganic cements act as binders.

Concrete is an example of a large class of particulate composites composed of nonmetallic particles in a nonmetallic matrix. A few of the other classes of particulate composites include (1) metals in metal, (2) metals in plastic, (3) ceramics and metal, (4) dispersion-hardened alloys, and (5) organic/plastic-inorganic.

Fillers

Fillers used in large quantities to reinforce plastics include alumina (aluminum oxide), calcium carbonate, calcium silicate, cellulose flock, cotton (different forms), short glass fiber, glass beads, glass spheres, graphite, iron oxide powder, mica, quartz, sisal, silicon carbide, titanium oxide, and tungsten carbide (Tables 15.21 and 15.22). The choice of filler varies and depends to a great extent on the requirements of the end product and method of fabrication.

Fillers offer a variety of benefits: increased strength and stiffness, reduced cost, shrinkage reduction, exothermic heat reduction, thermal expansion coefficient reduction, improved heat resistance, slightly improved heat conductivity, improved surface appearance, reduced porosity, improved wet strength, reduced crazing, improved fabrication mobility, increased viscosity, improved abrasion resistance, and impact strength. Fillers also have disadvantages. They may limit the method of fabrication, inhibit cure of certain resins, and shorten pot life of the resin.

Resin	Fillers
Amorphous ABS SAN Amorphous Nylon Polycarbonate Polystyrene Polysulfones	Lower tensile strength Can more than double flexural modulus Raise HDT slightly Embrittle resins Can impact special properties such as lubricity conductivity, flame retardance Reduce and balance shrinkage Reduce melt flow Can lower cost
Crystalline Aceals Nylon 6, 6/6, 6/10, 6/12, etc. Polypropylene Polyphenylene sulfide Polyesters Polyethylene	Lower tensile strength Can more than triple flexural modulus Raise HDT slightly Embrittle resins Can impart special properties such as lubricity, conductivity, magnetic properties, flame retardance Reduce shrinkage Reduce distortion Reduce melt flow Can lower cost

Table 15.21 Examples of fillers used in TP RPs (chapter 1)

Thermosets	Alumina	Calcium carbonate	Carbon black	Clay	Cotton flock	Glass bubbles	Glass fibers	Graphite	Mica	Quartz	Talc	Wollastonite Silicate	Wood flour
Diallyl o-phthalate		•	•	•		•	•				•		
Epoxy	•	•	•	•		•	•	•		•		•	
Phenolic	•	•	•	•	•	•	•	•	•		•	•	•
Polyester		•		•		•	•				•		•
Melamine					•		•		•		•	•	
Urea				•	•								•
Silicone	•			•			•	•		•			
Urethane	•	•	•	•		•	•			•		•	

Table 15.22 Examples of fillers used in TS RPs (chapter 1)

PROPERTIES

Fillers are strong, usually inert materials bound into a plastic to improve its properties such as strength, stiffness or modulus of elasticity, impact resistance, dimensional shrinkage, and so on (Tables 15.4, 15.6, 15.7, 15.23 to 15.40 and Fig. 15.25). They include fibers and other forms of material. There are inorganic and organic fibers that have the usual diameters ranging from about 1 to more than 100 micrometers. Properties differ for the different types, diameters, shapes, and lengths. To be effective, a reinforcement must form a strong adhesive bond with the plastic; for certain reinforcements, special cleaningand sizing, treatments are used to improve bonds.

A part's glass content has a great influence on its mechanical properties. In general, the more glass a part has the stronger it is. This occurs with the ability to pack more reinforcement (Fig. 15.26). Fiber content can be measured in percentage by weight of the fiber portion (wt%). However, it is also reported in percentage by volume (vol%) to reflect the better structural role of the fibers, which is related to volume (or area) rather than to weight (Figs. 15.26 and 15.27). When content is only in percentage, it usually refers to wt%.

The fiber content in mat- or random-fiber RPs is usually somewhat lower than for an isotropic laminate, which is composed of a number of unidirectional plies. Both laminates may, for example, be planar-isotropic. The random crisscross nature of chopped fibers in a mat does not permit close packing of the bundles, and thus the fiber content is usually low. With a layup of unidirectional plies, the packing of fibers within a ply may be very close, and the fiber content may be very high. The higher fiber content made from individual plies tends to make the product stiffer and stronger than a product made with the mat construction.

There is a relationship between the way the glass is arranged and the amount of glass that can be packed in a given product. By placing continuous strands, such as round glass fibers in a filament-winding pattern, next to each other in a parallel arrangement, more glass can be placed in a given volume (Fig. 15.26). Glass content can range from 65 wt% to 95.6 wt% or up to 90.8 vol%. When

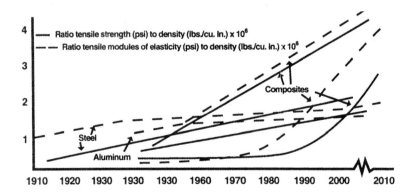

Table 15.23 Comparison of tensile properties in RPs, steel, and aluminum

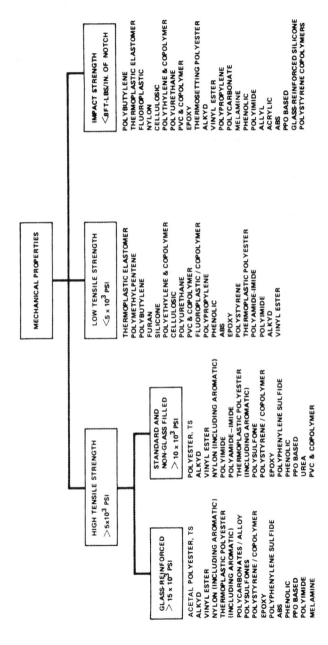

Table 15.24 Mechanical properties of resins that are reinforced to increase properties

Property	Physical					Mechanical				Thermal			Electrical		
	Specific gravity	Water absorption, 24 hrs.	Mold shrinkage	Tensile strength	Flexural modulus	Izod impact strength notched/unnotched	Rockwell hardness	Heat deflection temperature 264 psi	Long term[2] service temperature	Coefficient of linear thermal expansion	Flammability	Dielectric strength (ST)	Dielectric constant 60 Hz- 10^6 Hz	Dissipation factor 60 Hz- 10^6 Hz
Units	—	%	in./in.	psi	10^3 psi	ft.·lbs/in.	—	°F	°F	10^{-3} in./in./°F	UL Subj. 94	volts/mil.		
Acrylonitrile-butadiene-styrene (ABS)	1.28	0.14	0.001	14,500	1,100	1.4/6-7	M99.R124	220	155	1.6	HB	—	—	—
Styrene-acrylonitrile (SAN)	1.31	0.10	0.005-0.001	17,400	1,500	1.0/3-4	M94.R123	215	140	1.8	HB	—	—	—
Polystyrene	1.28	0.05	0.005-0.001	13,500	1,300	1.0/2-3	M92	215	120	1.9	HB	550	2.81-2.81	0.0007-0.0008
Styrenic copolymer	1.31	0.07	0.001	15,000	1,300	1.4/4-5	M96	250	140	1.8	HB	—	—	—
Polyethylene (HDPE)	1.17	0.015	0.0030	10,000	900	1.1/8-9	R85	260	185	2.7	HB	—	—	—
Polypropylene	1.13	0.03	0.004	9,800	800	1.65/5-6	M57.R111	295	220	2.0	HB	475	2.30-2.20	0.001-0.003
Polypropylene	1.13	0.02	0.0035	13,500	800	1.9/10-12	M57.R111	310	220	2.1	HB	—	—	—
Nylon-type 6	1.37	1.1	0.0035	23,000	1,200	2.3/20	M92.R121	420	215	1.7	HB	450	42.0-3.60	0.009-0.018
Nylon-type 6/12	1.30	0.20	0.0035	22,000	1,100	2.4/20	M93.R120	415	210	1.5	HB	440	4.20-3.50	0.013-0.015
Nylon-type 6/10	1.30	0.20	0.0035	21,000	1,100	2.4/20	M93.R120	420		1.5	HB	440	4.20-3.50	0.013-0.015
Nylon-type 6/6	1.37	0.9	0.004	26,000	1,300	2.0/17	M96.R121	490	240	1.8	HB	440	4.20-3.50	0.009-0.0180
Amorphous nylon	1.35	0.19	0.003	21,500	1,150	1.2/6-8	R119	285	185	1.8	HB	500	—	—
Super-tough nylon	1.30	0.60	0.004	17,000	900	4.0/20-22	R120	415	—	1.9	HB	525	3.95-3.95	0.0035-0.0065
Acetal	1.63	0.30	0.003	19,500	1,400	1.8/8-10	M86	300	220	2.2	HB	510	3.60-2.00	0.002-0.020
Thermoplastic polyester	1.52	0.06	0.003	20,000	1,200	2.6/16-18	M84.R119	430	280	1.2	HB	510	3.88-3.78	0.003-0.007
Polyphenylene sulfide	1.56	0.04	0.002	20,000	1,600	1.4/8-9	R121	500	355	1.3	VO+	500	3.71-3.61	0.0019-0.0043
Polyetheretherketone	1.49	0.11	0.003	22,800	1,500	1.8/15-16	—	600	482	1.2	VO+	500	3.50-3.43	0.0010-0.0075
Polycarbonate	1.43	0.07	0.002	18,500	1,200	3.7/17-18	M95.R118	300	260	1.3	V	480	3.55-3.49	0.0019-0.0049
Polysulfone	1.45	0.20	0.002-0.003	18,000	1,200	1.8/14	M92.1.108	365	300	1.4	VO	480	3.80-3.76	0.0030-0.004
Polyethersulfone	1.60	0.34	0.002	20,300	1,218	1.6/10	M98	420	374	1.3	VO	460	—	—
Thermoplastic polyurethane	1.46	0.25	0.004	8,200	190	9.5/28-29	D65	340	140	2.5	HB	—	—	—
Polyester elastomer	1.42	0.17	0.004	10,000	320	5.0/20	D70	340	150	2.5	HB	—	—	—
FEP	2.21	0.01	0.002-0.004	6,000	800	8.0/17	D63	350	390	2.4	VO	475	2.55-2.52	0.0020-0.002
ETFE	1.89	0.02	0.004-0.006	14,000	1,050	7.5/17-18	R74	460	350	—	VO	410	3.5-3.4	0.0006-0.005

Table 15.25 Properties per ASTM of 30 wt% glass-fiber RTPs

Resin[a]	Specific gravity	Tensile strength, MPa[b]	Tensile modulus, GPa[c]	Elongation, %	Flexural strength, MPa[b]	Flexural modulus, GPa[c]	Izod impact notched, J/m[d]	Deflection temperature under load, °C
Nylon-6,6								
Unreinforced	1.14	83	2.9	60	119	2.8	53	90
30% glass fibers	1.39	172	9.0	4	248	9.0	107	252
30% carbon fibers	1.28	227	20.7	3	324	20.7	85	263
40% mineral filler	1.50	92	5.5	3	155	7.2	48	249
40% glass-mineral	1.49	124	7.6	3	207	9.7	64	246
Polypropylene								
Unreinforced	0.9	34	1.4	11		1.6	51	53
30% glass fibers	1.13	52	5.5	2.5	65	4.1	64	137
30% glass fibers chemically coupled	1.13	83	5.9	2.3	110	5.5	107	151
40% mica	1.23	31	4.8	4	48	4.1	37	96
40% talc	1.25	29	3.1	4	48	3.1	27	76
Polycarbonate								
Unreinforced	1.20	65	2.4	7	93	2.3	801[e]	132
30% glass fibers	1.43	131	9.0	2.5	138	7.6	160	143
30% carbon fibers	1.33	152	17.2	1.8	220	15.2	107	149
5% stainless steel	1.27	68	3.1	5	110	3.1	69	146
Polyesters								
30% glass fibers (PBT)	1.51	121	6.9	4	200	8.7	96	206
30% glass fibers (PET)	1.56	158	8.7	3	234	9.1	107	224

[a] 6.4-mm thickness, except where otherwise stated.
[b] To convert MPa to psi, multiply by 145.
[c] To convert GPa to psi, multiply by 145,000.
[d] To convert J/m to ft-1bf/in., divide by 53.38.
[e] 3.2-mm thickness.

Table 15.26 Properties of glass-fiber RTPs with different glass fiber contents and other reinforcements

Property	Temperature, F	Short fiber		Long fiber	
		30%	50%	30%	50%
Tensile strength (10^3 psi)	300	12.8	13.8	14.3	19.2
Elongation (%)	300	9.3	7.8	5.3	5.6
Flexural strength (10^3 psi)	300	13.8	14.3	17.4	23.7
Flexural modulus (10^5 psi)	300	4.64	5.27	5.50	8.90
Tensile strength (10^3 psi)	400	6.3	7.3	7.8	8.3
Elongation (%)	400	8.6	9.5	6.2	6.8
Flexural strength (10^3 psi)	400	6.9	7.4	8.9	10.0
Flexural modulus (10^5 psi)	400	3.96	4.80	5.19	7.51

Table 15.27 Properties of short and long glass fiber-nylon 6/6 RPs at elevated temperatures

Property	How achieved	Sacrifice (from base resin) Amorphous	Sacrifice (from base resin) Crystalline	Comments
Increased tensile strength	Glass fibers Carbon fibers Fibrous materials	Ductility, cost Ductility, cost	Ductility, cost Ductility, cost Ductility	Glass fibers are the most cost effective way of gaining tensile strength. Reinforcement makes brittle resins tougher and embrittles tough resins.
Increased flexural modulus	Glass fibers Carbon fibers Rigid minerals	Ductility, cost Ductility, cost Ductility	Ductility, cost Ductility, cost Ductility	Any additive more rigid than the base resin produces a more rigid composite. Particulate fillers severely degrade impact strength.
Flame resistance	FR additive	Ductility, tensile strength, cost	Ductility, tensile strength, cost	FR additives interfere with the mechanical integrity of the polymer and often require reinforcement to salvage strength. They also narrow the molding latitude of the base resin. Some can cause mold corrosion.
Increased Heat-deflection temperature (HDT)	Glass fibers Carbon fibers Fibrous minerals	Ductility, cost Ductility, cost	Ductility, cost Ductility, cost Ductility	When reinforced, crystalline polymers yield much greater increases in HDT than do amorphous resins. As with tensile strength, fibrous minerals increase HDT only slightly. Fillers do not increase HDT.
Reduced coefficient of friction	PTFE Silicone MoSe Graphite	Cost	Cost	These fillers are soft and do not dramatically affect mechanical properties. PTFE loadings commonly range from 5 to 20%; the other are usually 5% or less. Higher loadings can cause mechanical degradation.
Reduced wear	Glass fibers Carbon fibers Lubricating additives	— — —	— — —	See chapter 22.
Electrical conductivity	Carbon fibers Carbon powders	Ductility, cost Tensile strength, ductility, cost	Ductility, cost Tensile strength, ductility, cost	Resistivities of 1 to 100,000 ohm-cm can be achieved and are proportional to cost. Various carbon fibers and powders are available with wide variations in conductivity yields in composites.

Table 15.28 Examples of obtaining desired properties of TP-RPs

Process	Tensile Strength			Tensile Modulus			Flexural Strength			Compressive Strength			Impact Strength			Thermal Conductivity				Heat Distortion at 1.8 MPa		Dielectric Strength	
	MPa	ksi		GPa		10^6 psi	MPa		ksi	MPa		ksi	J/m		ft · lbf/ft	W/m · K		Btu · in./h ft^2 · °F		°C	°F	kV/cm	kV/in.
Spray	60–120	9–18		5.5–12		0.8–1.8	110–190		16–28	100–170		15–25	210–640		48–144	0.17–0.23		1.2–1.6		175–205	350–400	80–160	200–400
Compression	170–210	25–30		6.2–14		0.9–2.0	70–280		10–40	100–210		15–30	530–1,050		120–240	0.19–0.26		1.3–1.8		175–205	350–400	120–240	300–600
Filament winding	550–1,700	80–250		28–62		4.0–9.0	690–1,850		100–270	310–480		45–70	2,150–3,200		480–720	0.27–0.33		1.9–2.3		175–205	350–400	120–160	300–400
Pultrusion	410–1,050	60–150		28–41		4.0–6.0	690–1,050		100–150	210–480		30–70	2,400–3,200		540–720	0.27–0.33		1.9–2.3		205–260	400–500	80–160	200–400

Table 15.29 Properties of RPs with 30 wt% to 50 wt% glass fiber-TS polyester based on fabricating process

Polyester: Sheet Molding Compounds

Resin type		Typical values			
Reinforcement/structure		20% glass	28% glass	30% glass	50% glass
US figures are given in tinted panels					
Glass transition temperature	°C		140–200		
	°F		280–390		
Density/specific gravity	g/cm³	1.78	1.84	1.85	2.0
	lb/in³	0.064	0.66	0.066	0.072
Mechanical properties					
Tensile strength	MPa	37	61.9	84	161
	10³ psi	5.3	9.0	12.0	23.0
Tensile modulus	GPa	10.6		13.0	17.2
	10⁶ psi	1.4		1.70	2.27
Elongation at break	%	0.4		1.7	
Bending/flexural strength	MPa	112	176.8	182	315
	10³ psi	16.0	25.7	26.0	45.0
Bending/flexural modulus	GPa	10.6	12.9	12.1	15.2
	10⁶ psi	1.4	1.7	1.6	2.0
Compressive strength	MPa	161		168	224
	10³ psi	23.0		24.0	32.0
Izod impact strength	ft lbs/in	8.20	15.0	16.0	19.4
Thermal properties					
Coefficient of linear thermal expansion	k⁻¹				
	x10⁻⁵ in/in/°F				9.4
Thermal conductivity	W/mk				
	btu(h.ft²/°F)				
Heat deflection temperature	°C x 1.8 MPa	>204		>204	>204
	°F x 264 psi	>400		>400	>400
Continuous use temperature	°C				
	°F				
Flammability properties					
Spread of flame	BS 476				
Burn rate (ISO 3795)	mm/min				
Limiting Oxygen Ind	%				
UL flammability	1.3–10mm	5V		5V	5V
Electrical properties					
Tracking resistance	Volts				
Comp. tracking resist	CTI				
Surface resistivity	Log₁₀ Ohm				
	Ohms/sq				
Volume resistivity	Ohms-cm			5.7 × 10¹⁴	
Dissipation factor	1MHz				
Dimensional change					
Water absorption	%	0.10	0.25	0.25	0.50
Mold shrinkage	in/in	0.002			

Notes: Hardness = Barcol 68.

Table 15.30 Properties of TS polyester RPs with different amounts of glass fibers

Polyester: Sheet Molding Compounds

Resin type		Low shrink (0.1%)	Low profile/ zero shrink	Low profile high perf.
Reinforcement/ structure				
US figures are given in tinted panels				
Glass transition temperature	°C	190	210	170–200
	°F	375	410	340–390
Density/ specific gravity	g/cm^3			
	lb/in^3			
Mechanical properties				
Tensile strength	MPa	95	95	85–95
	10^3 psi	13.6	13.6	12.1–13.6
Tensile modulus	GPa	12.0	12.5	10.0–10.5
	10^6 psi	1.57	1.64	1.31–1.38
Elongation at break	%	1.9	1.6	1.7–1.8
Bending/ flexural strength	MPa	200	200	170–190
	10^3 psi	28.5	28.5	24.3–27.1
Bending/ flexural modulus	GPa	14.5	14.5	10.5–12.0
	10^6 psi	1.90	1.90	1.38–1.58
Compressive strength	MPa			
	10^3 psi			
Izod impact strength	Charpy kJ/m^2	110	120	110–120
	ft lbs/in			
Thermal properties				
Coefficient of linear thermal expansion	k^{-1}	23	18	18
	x10^{-5} in/in/°F			
Thermal conductivity	W/mk			
	btu(h.ft^2/°F)			
Heat deflection temperature	°C x 1.8 MPa			
	°F x 264 psi			
Continuous use temperature	°C			
	°F			
Flammability properties				
Spread of flame	BS 476			
Burn rate (ISO 3795)	mm/min			
Limiting Oxygen Ind	%			
UL flammability	1.3–10mm			
Electrical properties				
Tracking resistance	Volts			
Comp. tracking resist	CTI			
Surface resistivity	Log$_{10}$ Ohm Ohms/sq			
Volume resistivity	Ohms-cm			
Dissipation factor	1MHz			
Dimensional change Water absorption	%			
Mold shrinkage	in/in			

Notes: Low profile SMCs can give LORIA waviness index reading of about 50:1.5–20 long waviness by Daimler Benz/ BASF method ($\lambda = 10 - 100$ mm).

Table 15.31 Properties of glass fiber mats RPs with different types of TS polyesters

Material		Glass fiber, Wt.%	Specific gravity	Thermal coefficient of expansion	Heat deflection at 1.8 MPa, ·C° (°F)		Thermal conductivity, (W/m·K) (BTU·in./ hr.ft.² °F)		Specific heat, J/(kg·K)	Tensile strength, MPa (psi)		Tensile modulus, GPa (kip/in.²)	
			D 792	D 696	D 648		C177			D638		D638	
Glass-fiber-reinforced thermosets (RTS)	Polyester SMC, compression	30.0	1.85		200+	(392+)			1.26	83	(12,000)	11.7	(1,700)
	Polyester SMC, compression	20.0	1.78		200+	(392+)			1.26	36.5	(5,300)	11.7	(1,700)
	Polyester SMC, compression	50.0	2.00	9.4	200+	(392+)			1.26	158	(22,900)	15.7	(2,280)
	Polyester BMC, compression	22.0	1.82	6.6	260	(500)	8.37	(58.1)	1.26	41.3	(5,990)	12.1	(1,750)
	Polyester BMC, injection	22.0	1.82	6.6	260	(500)	8.37	(58.1)	1.26	33.5	(4,860)	10.5	(1,520)
	Epoxy filament wound	80.0	2.08	2.0	200+	(392+)	1.77	(12.3)	0.96	552	(80,000)	27.6	(4,000)
	Polyester, pultruded	55.0	1.69	5.0			6.92	(48.0)	1.17	207	(30,000)	17.2	(2,490)
	Polyurethane, milled fibers (RRIM)	13.0	1.07	78.0	29	(84)				19.3	(2,800)		
	Polyurethane, flaked glass (RRIM)	23.0	1.17	53.1						30.4	(4.410)		
	Polyester spraying/lay-up	30.0	1.37	12.0	200+	(392+)	2.60	(18.0)	1.30	86.2	(12.500)	6.9	(1,000)
	Polyester, woven roving, lay-up	50.0	1.64	4.0	200+	(392+)				255	(37,000)	15.5	(2,250)

Table 15.32 General properties of TS RPs per ASTM testing procedures

Elongation, %	Flexural modulus, GPa (kip/in^2)	
D 638	D 790	
<1.0	11.0	(1,600)
0.4	9.7	(1,400)
1.7	13.8	(2,000)
0.5	10.9	(1,580)
0.5	9.9	(1,400)
1.6	34.5	(5,000)
	11.0	(1,600)
140.0	0.26–0.37	(38–54)
38.9	1.0	(145)
1.3	5.2	(7,500)
1.6	15.5	(2,250)

Compressive strength, MPa (psi)	Impact strength Izod at 22°C, J/m	Hardness	Water absorption in 24 hr. %	Mold shrinkage, %
D 695	D 256	D 785	D 570	D 955
166 (24,100)	854	Barcol 68	0.25	
159 (23,100)	438	Barcol 68	0.10	0.002
221 (32,000)	1,036	Barcol 68	0.50	
138 (20,000)	227	Barcol 68	0.20	0.001
	154	Barcol 68	0.20	0.004
310 (45,000)	2,400	M98	0.50	0.008
207 (30,000)	1,335	Barcol 50	0.75	
		Shore D 65–75		
	112			
152 (22,000)	690–800	Barcol 50	1.30	

Table 15.32 General properties of TS RPs per ASTM testing procedures *(continued)*

	Room temperature tensile strength (10^3 psi)	Room temperature compressive strength (10^3 psi)	Ultimate flexural strength (10^3 psi)				Flexural modulus (10^3 psi)		
			At room temperature	After 30-day water immersion	At 300°F	At 500°F	At room temperature	At 300°F	At 500°F
Fabric Laminates									
Epoxy-glass									
Bisphenol A	50	55	85–89		40–60	8–11	3600–3800	2900–3300	3200
Novalac	42	57	85–90		58–63	25	3200–3400	2700	
Peracetic acid	57–59	49–53	85–90		60–65	47	4000–4200	3700–3900	
Bisphenol A and high-modulus glass	64	61	90	76	44		5000		
High-temperature epoxy and high-modulus glass	55	59	86			23	5300	3700	3000
Phenolic-glass									
Standard grade	50	35	50			40	3000		3000
Heat-resistant	65	48	86	52	78	47–52	4800	4000	3300–3600
Silane-modified	46	51	80	73	65	65	4300	3500	3500
Phenolic-leached glass	14	24	25		20		2600	2600	2600
Phenolic-quartz	35		68		37	50	3700	3400	3200
Polyesters									
Styrene-modified	50–55	40–45	50–75	64	30	24	3300	2500	2000
TAC-modified	59	51	65	59	43	30	3000	2000	1800
Styrene-modified and high-modulus glass		52	59	49			4200		
Silicone-glass	41	29	50	36	23	21	2900	2400	2200
Filament-Wound Structures									
Epoxy-glass	270	70	271		245				
Epoxy and high-tensile glass	285								
Mil P-27327 (proposed)	200	50	220	210	200	180	7000	6800	6800
Molding Compounds									
Phenolic-glass	10	35	20		16.4		3300		
Phenolic-leached-glass	8.5	36	14				3000		
Phenolic-asbestos	9	24	24			15	2000		1400
Silicone-glass	4.5	16	14		6	6	2100	1100	500
Polyester-glass	9.5	29.5	19		12.7	6.7	1400	910	910

Table 15.33 Examples of mechanical properties of TS RPs at ambient and elevated temperatures

	Flexural modulus (10^6 psi)		
Test conditions	Polyester	Epoxy	Phenolic
Atmospheric pressure and room temperature	4.67	3.54	4.17
Low pressure (10^{-6} torr) at room temperature	4.67	3.54	4.17
Low pressure (10^{-6} torr), ultraviolet (0.036 W/cm^2) and moderate temperature (250°F)	3.84	2.14	3.94
Atmospheric pressure, ultraviolet (0.036 W/cm^2) and moderate temperature (250°F)	3.38	2.00	3.86
Atmospheric pressure, ultraviolet (0.036 W/cm^2) and room temperature	3.75	3.54	3.68
Low pressure (10^{-6} torr) and moderate temperature (250°F)	3.61	2.38	3.96
Atmospheric pressure and moderate temperature (250°F)	3.75	2.25	3.81

Table 15.34 Flexural modulus of glass-polyester-RPs exposed to various environmental elements

half of the strands are placed at right angles to each half, glass loadings range from 55 wt% to 88.8 wt% or up to 78.5 vol%.

Types of reinforcements include fibers of glass, carbon, graphite, boron, nylon, polyethylene, PP, cotton, sisal, asbestos, metals, whiskers, and so on. Other types and forms of reinforcements include bamboo, burlap, carbon black, platelet forms (including mica, glass, and aluminum), fabric, and hemp. The so-called whiskers are metallic or nonmetallic single crystals (micrometer-size diameters) of ultrahigh strength and modulus. Their extremely high performances (high modulus of elasticity, high melting points, resistance to oxidation, low weights, etc.) are attributed to their near-perfect crystal structure, chemically pure nature, and fine diameters that minimize defects. They exhibit a much higher resistance to fracture (toughness) than other types of reinforcing fibers (12).

In general, adding reinforcing fibers significantly increases mechanical properties. Particulate fillers of various types usually increase the modulus, plasticizers generally decrease the modulus but enhance flexibility, and so on.

It can generally be claimed that fiber-based RPs offer good potential for achieving high structural efficiency coupled with a weight savings in products, fuel efficiency in manufacturing, and cost effectiveness during service life. Conversely, special problems can arise from the use of RPs, due to the extreme anisotropy of some of them, the fact that the strength of certain constituent fibers is intrinsically variable, and because the test methods for measuring an RP's performance need special consideration if they are to provide meaningful values.

Some of the advantages, in terms of high strength-to-weight ratios and high stiffness-to-weight ratios, have been presented. This shows that some RPs can outperform steel and aluminum if those metals are in their ordinary forms. If bonding to the matrix is good, then fibers augment mechanical

	Unit	Single skin CSM/WR polyester	Single skin aramid/ glass epoxy	Sandwich aramid/ glass epoxy	Sandwich carbon/ aramid epoxy	Strip wood/ epoxy	Framed aluminum alloy	Framed steel
Density	kg/m³	1900	1300	1400	2010	1800	1350	1450
Fiber weight fraction %	%	52	46	47	60	58	50	58
0° Tensile strength	MPa	352	360	424	1050	1090	1000	1133
0° Tensile modulus	GPa	13.46	21.97	31.38	55.0	67.0	70.0	100.0
Ultimate strain	%	2.7	1.7	1.4	3.3	2.3	1.8	1.28
0° Compressive strength	MPa	242	75	122	350	360	200	380
0° Compressive modulus	GPa	12.99	16.27	19.07	47.0	57.0	50.0	80.0
90° Tensile strength	MPa	283	310	389	39	39	37	42
90° Tensile modulus	GPa	11.70	18.57	32.98	6.0	6.0	5.7	5.9
90° Compressive strength	MPa	242	75	122	100	95	90	95
90° Compressive modulus	GPa	12.99	14.05	19.07	9.0	10.0	9.0	11.0
Shear strength	MPa	55	38	50	55	55	40	58
Shear modulus	MPa	5.0	5.0	5.0	4.7	4.75	2.57	4.8

Table 15.35 Strength and modulus for glass fiber-TS RPs at low temperature

Table 15.36 Coefficients of thermal expansion for parallel glass fiber-TS RPs

Resin type	Temperature, °F	Coefficient of thermal expansion			
		Parallel to warp, 10^6 in. per in.	Perpendicular to warp, 10^6 in. per in.	45° to warp, 10^6 in. per in.	Through thickness, 10^6 in. per in.
Epoxy-phenolic	−100 to 200	4.8	5.0	5.0	10.0
	300 to 600	2.8	2.5	4.5	6.3
Silicone (MIL-R-25506)	−100 to 100	4.0	5.0	5.0	38.0
	100 to 600	3.0	3.0	3.0	80.0
Phenolic [a] (MIL-R-9299)	−100 to 200	6.0	5.8	6.4	11.1
	300 to 600	3.2	2.9	3.0	6.2
Polyester [b] (MIL-R-7575)	−100 to 100	7.8	9.3	8.5	19.1
	200 to 400	1.4	2.3	1.3	237.6
Triallyl cyanurate polyester (MIL-R-25402)	−100 to 200	5.5	5.2	5.2	11.0
	350 to 600	3.6	3.9	3.6	12.0
Epoxy (MIL-R-9300)	−100 to 200	5.5	6.7	6.7	−
	300 to 600	3.3	1.5	2.3	−

[a] Average of data from laminates made with 181 fabric and two mats
[b] Average of data from laminates made with several styles of fabric (116, 112, 181 and 143)

Class	Material types	Limited dose for sealing properties (rad)	Continuous service temperature range (°F)	Remarks
I	Phenolic resin – glass fiber Epoxy resin – glass fiber Modified epoxy – phenolic – glass fiber (heat resistant epoxy)	10^{10} 5×10^9 5×10^9	−423 to 500 −423 to 250	Preferred for general use in interior and exterior structural applications particularly where moderate to high temperatures and radiation exposures are encountered; excellent mechanical strength properties and good stability to vacuum and UV radiation
II	Polyester resin – glass fiber Melamine resin – glass fiber Silicone resin – glass fiber Triallylcyanurate resin (TAC) – glass fiber (heat resistant polyester)	10^9 10^9 10^9	−423 to 250 −423 to 500 −423 to 500	Not preferred for general primary structural use because of lower strength properties, and slightly poorer stability to the space environment; recommended for external or internal electrical applications where optimum dielectric properties are required (e.g., radomes) particularly at moderate to high temperatures and moderate radiation exposures
III	Phenolic, polyester, epoxy resins and modifications filled or reinforced with organic fibers (e.g., Dacron and Orlon)	10^8	−100 to 250	Not recommended for structural applications but may be used in certain non-structural internal applications such as dielectrics. These materials may be used instead of I and II only under exceptional circumstances, after thorough review of design and environment application. Relatively low mechanical strength properties and temperature and radiation stability. Good vacuum and dimensional stability. Good electrical properties

Many combinations of resins and reinforcement types and weaves available for specific structural applications. Directional strength properties can be varied from unidirectional to orthotropic by choice of reinforcement type and laminate fabrication method.

Other fibrous inorganic reinforcements, e.g., Refrasil quartz or asbestos; should be equally suitable for use in space but data lacking to support recommending them. These reinforcements are generally used for more specialized applications such as thermal insulation, ablation, etc., where mechanical strength properties are secondary to heat resistance.

Table 15.37 Example of TS RPs for electrical applications

Material	Ultraviolet intensity[b] (W/cm²)	Exposure time (hr)	Maximum temperature reached (°F)	Weight loss[c] (%)	Average[d] ultimate flexure strength (psi)	Average[a] flexural modulus (10⁶ psi)	Average[a] compressive strength (psi)	Average[a] compressive modulus (10⁶ psi)
Polyester (P-43)	0						40,900	3.1
	0.036	125	250	0.7	59,300	2.4	47,300	3.1
	0.054	125	290	1.1	61,800	2.6	51,900	3.2
	0.072	25	325	2.5	64,400	2.8	40,800	3.1
	0.090	3	338	3.6	50,100	2.5	39,900	3.1
	0.108	3	442	5.1	24,200			
Epoxy (Epon 815)	0						53,900	3.5
	0.036	125	270	0.5	84,400	3.8	48,400	3.5
	0.054	125	300	0.8	84,700	3.8	51,300	3.5
	0.072	25	335	1.7	84,000	3.8	43,800	3.4
	0.090	3	342	2.1	57,600	3.5	44,200	3.4
	0.108	3	448	2.3	39,800			
Phenolic (91-LD)	0						44,800	3.5
	0.036	125	275	0.5	68,900	3.6	39,400	3.5
	0.054	125	320	1.2	61,700	3.6	39,300	3.5
	0.072	25	350	1.3	56,700	3.5	31,500	3.4
	0.090	3	402	1.6	49,500	3.0	32,600	3.4
	0.108	3	460	1.5	57,100	3.2		

[a] Pressure range during exposures: 7.1×10^{-5} (at highest intensity) to 6.0×10^{-6} torr.
[b] Wavelength range: 2000–4200 Å (0.0164 W/cm² = 1 sun in this wavelength range).
[c] Specimens exposed to high vacuum at 70°F for 1000 hr had negligible weight loss (< 0.1%).
[d] Average of four specimens.

Table 15.38 Mechanical properties of glass fabric-TS polyester RPs exposed to various intensities of near-UV radiation in a vacuum

Resin type	Test	Exposure (rad)	Temperature (°F)	Exposure time (hr)	Ultimate strength (psi)	Flexural modulus (10⁶ psi)
Silicone[a]	Flexure	None	Room	None	31,760	3.06
		8.3 × 10⁷	Room	200	31,460	2.94
		None	500	50	12,390	1.90
		2.1 × 10⁷	500	50	13,625	2.0
		None	500	100	13,410	2.0
		4.15 × 10⁷	500	100	11,720	2.0
		None	500	200	14,060	2.0
		8.3 × 10⁷	500	200	9,860	1.9
Heat-resistant epoxy[b]	Compression	None	Room	None	46,680	
		8.3 × 10⁷	Room	200	46,660	
		None	500	50	3,705	
		2.1 × 10⁷	500	50	3,780	
		None	500	100	4,090	
		4.15 × 10⁷	500	100	5,490	
		None	500	200	4,720	
		8.3 × 10⁷	500	200	6,360	
Phenolic[b]	Flexure	None	Room	None	84,525	4.22
		8.3 × 10⁷	Room	200	84,040	4.35
		None	500	50	27,300	3.14
		2.1 × 10⁷	500	50	55,020	3.46
		None	500	100	17,660	2.62
		4.15 × 10⁷	500	100	47,015	3.61
		None	500	200	12,330	2.13
		8.3 × 10⁷	500	200	15,645	2.41

[a] 181 glass fabric (112 finish).
[b] 181 glass fabric (Volan-A finish).

Table 15.39 Mechanical properties of glass fiber fabric-TS polyester RPs after irradiation at elevated temperatures

strength by accepting strains transferred from the matrix, which otherwise would break. This occurs until catastrophic debonding takes place. Particularly effective here are combinations of fibers with plastic matrices, which often complement one another's properties, yielding products with acceptable toughness, reduced thermal expansion, low ductility, and a high modulus.

ORIENTATION OF REINFORCEMENT

Both plastics and fibers influence orientation properties. For example, with certain TPs (including liquid crystal polymer [LCPs]), the plastic's molecular orientation can be used to aid in increasing

Material	Strength (psi) (MPa) (Yield values except where noted)			Modulus of Elasticity (E)	Coefficient of Thermal Expansion
	Tension	Compression	Shear	(ksi) (GPa)	(°F^{-1}) (10^{-6})
Wood (dry) parallel-to-grain					
Douglas fir	6,000 (41.4)	3,500 (24.1)	500 (3.45)	1,700 (11.7)	2
Redwood	6,500 (44.8)	4,500 (31.0)	450 (3.10)	1,300 (8.96)	2
Southern pine	8,500 (58.6)	5,000 (34.5)	600 (4.14)	1,700 (11.7)	3
Steel					
Mild, low-carbon	36,000 (248)	36,000 (248)	20,000 (138)	29,000 (200)	6.5
Cable, high strength	275,000* (1,896)	—	—	25,000 (172)	6.5
Concrete					
Stone	200* (1.38)	3,500 (24.1)	180* (1.24)	3,500 (24.1)	5.5
Structural, lightweight	150* (1.03)	3,500 (24.1)	130* (0.80)	2,100 (14.5)	5.5
Brick masonry	300* (2.07)	4,500 (31.0)	300* (2.07)	4,500 (31.0)	3.4
Aluminum, structural	30,000 (207)	30,000 (207)	18,000 (124)	10,000 (69.0)	12.8
Iron, cast	20,000* (138)	85,000 (586)	25,000* (172)	25,000 (172)	6
Glass, plate	10,000* (69.0)	36,000 (248)	—	10,000 (69.0)	4.5

Table 15.40 Properties of different materials

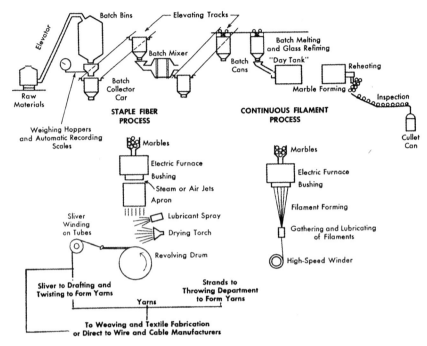

Figure 15.25 Staple glass fiber and continuous glass filament fiber process methods.

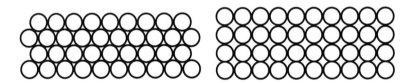

Figure 15.26 Fiber arrangements and property behavior (courtesy of Plastics FALLO).

stiffness, strength, and toughness, as well as craze and microcrack resistance in the direction of the plane or the plane of orientation (chapter 1). A source of orientation is fibers that cause positive and significant increased performance.

These effects are obvious in continuous filament winding (FW). Not so obvious are the anisotropic materials properties resulting from many TP processes. Viscous melt flow of hot plastic into injection molds produces an oriented structure (Fig. 4.26), usually having greater strength with crystalline TPs in the direction of flow than perpendicular to this direction. Shrinkage is usually greatest with crystalline TPs perpendicular to the direction of flow. With amorphous TPs, the greatest shrinkage can be in the direction of flow; however, as in IM, melt flow control can

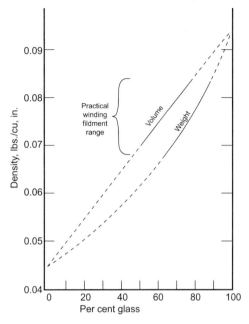

Figure 15.27 RP density versus percentage glass by weight or volume.

readily have even shrinkage in all directions. With fibers, shrinkage is less than unreinforced RPs (URPs) in the flow and perpendicular directions. Fig. 15.28 relates fiber directional properties to processes.

Flow within molds in processes such as IM and resin-transfer molding (RTM), or through dies in the extrusion process, can orient the molecules of the resin matrix and short or long fibers. This orientation can result in the desired design properties or, if not properly processed, can result in inferior properties that may become evident in the form of reduced resistance to crazing, low impact strength, lowered creep-rupture strength, and so on.

The behavior of RPs is dominated by the arrangement and the interaction of the stiff, strong fibers with the less stiff, weaker plastic matrix. A major advantage is the fact that directional properties can be maximized. As reviewed in chapter 19, they can be isotropic, bidirectional, orthotropic, and so on (Fig. 15.29 and Tables 15.41 and 15.42). Woven fabrics that are generally directional in the 0° and 90° angles contribute to the mechanical strength at those angles (Fig. 15.30). The rotation of alternate layers of fabric to a layup of 0°, +45°, 90°, and −45° alignment reduces maximum properties in the primary directions, but increases in the +45° and −45° directions. Different fabric weave patterns and/or individual fiber patterns are used to develop different property performances.

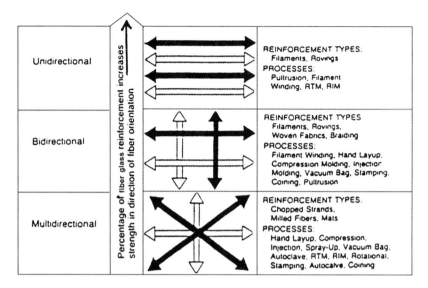

Figure 15.28 Fiber orientation provides different directional properties.

DIRECTIONAL PROPERTY

Different constructions and layups regarding directional properties are used with RPs. The next sections describe these constructions.

ANISOTROPIC

With this construction, the properties are different depending on the directions along the laminate flat plane. Anisotropic identifies a material that exhibits different properties in response to stresses applied in different directions along the axes.

BALANCED

This is a woven construction where equal parts of warp and fill fibers exist. It is the construction in which reactions to tension and compression loads result only in extension or compression deformations; in flexural loads it produces pure bending of equal magnitude in axial and lateral directions. It is an RP in which all laminae at angles other than 0° and 9° occur in ± pairs (not necessarily adjacent) and are symmetrical around the central line.

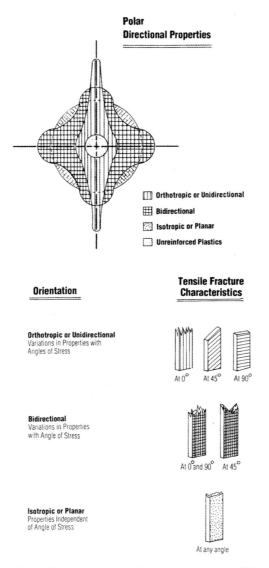

Figure 15.29 Examples of how fiber orientation influences properties of RPs.

Fiber	Axial modulus GPa (Msi)	Transverse modulus GPa (Msi)	Inplane shear modulus GPa (Msi)	Poisson's Ratio	Axial tensile strength MPa (Ksi)	Transverse tensile strength MPa (Ksi)	Axial compressive strength MPa (Ksi)	Transverse compressive strength MPa (Ksi)	Inplane shear strength MPa (Ksi)
E-glass	45 (6.5)	2 (1.8)	5.5 (0.8)	0.28	1020 (150)	40 (7)	620 (90)	140 (20)	70 (10)
Aramid	76 (11)	5.5 (0.8)	2.1 (0.3)	0.34	1240 (180)	30 (4.3)	280 (40)	140 (20)	60 (9)
Boron	210 (30)	19 (2.7)	4.8 (0.7)	0.25	1240 (180)	70 (10)	3310 (480)	280 (40)	90 (13)
SM carbon (PAN)	145 (21)	10 (1.5)	4.1 (0.6)	0.25	1520 (220)	41 (6)	1380 (200)	170 (25)	80 (12)
UHS carbon (PAN)	170 (25)	10 (1.5)	4.1 (0.6)	0.25	3530 (510)	41 (6)	1380 (200)	170 (25)	80 (12)
UHM carbon (PAN)	310 (45)	9 (1.3)	4.1 (0.6)	0.20	1380 (200)	41 (6)	760 (110)	170 (25)	80 (12)
UHM carbon (pitch)	480 (70)	9 (1.3)	4.1 (0.6)	0.25	900 (130)	20 (3)	280 (40)	100 (15)	41 (6)
UHK carbon (pitch)	480 (70)	9 (1.3)	4.1 (0.6)	0.25	900 (130)	20 (3)	280 (40)	100 (15)	41 (6)

Table 15.41 Properties of unidirectional RPs using different types of fibers

Property	Units	Graphite-epoxy		Graphite-epoxy RC = 29–30% VC = 1.7–2.4%		Graphite-polyimide RC = 35% VC = 0%		Graphite-polyimide RC = 27.5–31% VC = 0%		Graphite-polysulfone RC = 33–34% VC = 0–1.9%	
		22°C (72°F)	277°C (350°F)	22°C (72°F)	177°C (350°F)	22°C (72°F)	117°C (350°F)	22°C (72°F)	177°C (350°F)	22°C (72°F)	117°C (350°F)
Unidirectional laminate											
Longitudinal (0°) properties											
Tensile strength	Ksi(MPa)	218 (1502)	208 (1433)	197.7 (1362)	141.7 (976)	156.7 (1080)	152.2 (1049)	203.3 (1401)	187.4 (1291)	187.9 (1295)	179.1 (1234)
Tensile modulus of elasticity	Msi(GPa)	26.3 (181)	28.5 (196)	20.3 (140)	19.3 (133)	20.25 (140)	19.56 (135)	18.3 (126)	18.9 (130)	16.3 (112)	17.5 (121)
Compressive strength	Ksi(MPa)	218 (1502)	206 (1419)	157.4 (1084)	148.1 (1020)	180.0 (1240)	120.0 (827)	206.1 (1420)	164.6 (1134)	102.1 (703)	90.2 (621)
Compressive modulus of elasticity	Msi(GPa)	23.0 (158)	22.5 (155)	19.8 (136)	25.0 (172)	18.2 (125)	20.5 (141)	18.7 (129)	19.1 (132)	17.3 (119)	18.5 (127)
Flexural strength	Ksi(MPa)	247 (1702)	196 (1350)	200.7 (1383)	96.4 (664)	204.0 (1406)	179.1 (1234)	224.4 (1546)	178.8 (1232)	191.5 (1319)	135.2 (932)
Flexural modulus of elasticity	Msi(GPa)	—	—	17.77 (122)	16.29 (112)	16.89 (116)	18.43 (127)	18.4 (127)	17.3 (119)	17.8 (123)	20.0 (138)
Interlaminar shear strength (short beam)	Ksi(MPa)	15.9 (110)	8.9 (61)	12.69 (87.4)	7.18 (49.5)	14.8 (102)	10.2 (70.3)	13.62 (93.8)	9.79 (67.5)	11.6 (79.9)	8.4 (57.9)
Transverse (90°) properties											
Tensile strength	Ksi(MPa)	3.85 (26.5)	2.89 (19.9)	4.9 (33.8)	3.7 (25.5)	2.82 (19.4)	2.97 (20.5)	5.37 (37)	3.81 (26.3)	5.02 (34.6)	5.39 (39.1)
Tensile modulus of elasticity	Msi(GPa)	1.50 (10.3)	1.78 (12.3)	1.3 (9.0)	1.05 (7.2)	1.50 (10.3)	1.15 (7.9)	1.39 (9.6)	1.09 (7.5)	1.15 (7.9)	1.07 (7.37)

RC = resin content by weight.
VC = void content by volume.

Table 15.42 Properties of unidirectional graphite fiber-thermoplastic RPs varying in resin content by weight and varying in void content by volume (at 72°F and 350°F)

Biaxial load

Here is a loading condition in which a product is stressed in two different directions in its plane, such as a loading condition of a pressure vessel under internal pressure and with unrestrained ends.

Bidirectional

This construction has fibers oriented in various directions in the plane of the laminate, usually a cross laminate with the fibers 90° apart.

Isotropic

This construction has uniform properties in all directions. The measured properties of an isotropic material are independent on the axis of testing. The material will react consistently even if stress is applied in different directions; the stress-and-strain ratio is uniform throughout the flat plane of the material.

Isotropic transverse

This construction refers to a material that exhibits a special case of orthotropy in which properties are identical in two orthotropic dimensions but not the third. It has identical properties in both transverse directions but not in the longitudinal direction.

Nonisotropic

This is a material or product that is not isotropic; it does not have uniform properties in all directions.

Orthotropic

This construction has three mutually perpendicular planes of elastic symmetry.

Quasi-isotropic

This construction approximates an isotropic layout by orientation of plies in several or more directions.

Unidirectional

This construction refers to fibers that are oriented in the same direction, such as unidirectional fabric, tape, or laminate; this arrangement is often called *UD*. Such parallel alignment is included in pultrusion and filament-winding applications.

Z-AXIS

This construction is the reference axis normal (perpendicular) to the *X-Y* plane (the so-called flat plane) of the RP.

HETERGENEOUS/HOMOGENEOUS/ANISOTROPIC

Heterogeneous RPs have properties that vary, so the composition varies from section to section in a heterogeneous mass that has uniform properties. For design purposes, many heterogeneous materials are treated as homogeneous (uniform). This is because a reasonably small sample of material cut from anywhere in the body has the same properties as the body. An unfilled (unreinforced) TP is an example of this type of material.

The designer must be aware that as the degree of anisotropy increases, the number of constants or moduli required to describe the material increases. With isotropic construction, one could use the usual independent constants, namely Young's modulus and Poisson's ratio, to describe the mechanical response of materials. RPs are either constructed from a single layer or built up from multiple layers. The properties of each layer are usually orthotropic, which is a special case of anisotropy. Fibers that remain straight in the single layer are desired. However, with many fabrics, they are woven into configurations that kink the fiber bundles severely. Such fabric constructions may be very practical since they drape better over doubly warped molds than do fabrics that contain predominantly straight fibers.

There are fiber bundles in lower-cost woven roving that are convoluted or kinked as the bulky rovings conform to a square weave pattern. Kinks produce repetitive variations in the direction of reinforcement with some sacrifice in properties. Kinks can also induce high local stresses and early failure as the fibers try to straighten within the matrix under a tensile load. Kinks also encourage local buckling of fiber bundles in compression and reduce compressive strength. These effects are particularly noticeable in tests with woven roving, in which the weave results in large-scale convolutions. Regardless, extensive use of fabrics is made based on their capabilities. Examples of properties for different E-glass fabric constructions and layups with TS polyester plastic moldings are shown in Figures 15.30 to 15.34.

MATERIAL OF CONSTRUCTION

Reinforcements can be in continuous forms (fibers, filaments, woven or nonwoven fabrics, tapes, rovings, etc.), chopped forms of different lengths (Fig. 15.17), or discontinuous forms (whiskers, flakes, spheres, etc.). The reinforcements can allow the RP materials to be tailored to the design or the design to be tailored to the material (383).

At least 80 wt% of glass fiber used in RPs is the E-type. E-types are suitable for electrical grades, so they are also called electric glass (borosilicate type). There are also C-types for chemical stability, D-types for precisely controlled dielectric constants (high boron content), and M-types

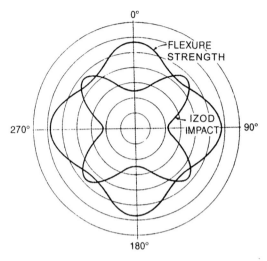

Figure 15.30 Parallel/bidirectional layup of woven fabric 181 glass fiber (courtesy of Plastics FALLO).

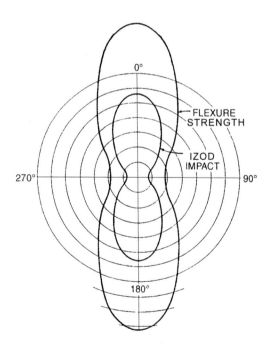

Figure 15.31 Parallel/unidirectional layup woven fabric 143 glass fiber (courtesy of Plastics FALLO).

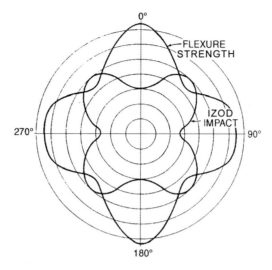

Figure 15.32 Ply layup at 0° and 90° woven fabric 143 glass fiber construction (courtesy of Plastics FALLO).

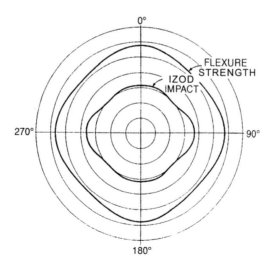

Figure 15.33 Ply layup at 0°, 45°, 90°, and 135° woven fabric 143 glass fiber construction (courtesy of Plastics FALLO).

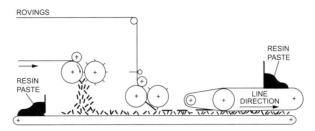

Figure 15.34 Sheet molding compound (SMC) production line using chopped glass fiber including roving to provide bidirectional properties, cutting continuous rovings for ease of mold-cavity fit.

for high modulus of elasticity (high berlia content). R-types are a cross between E- and S-types with limited production; S-types have high tensile strength (magnesia-alumina-silicate content).

Fibers should go through binder/sizing coupling agent treatments to maximize their performances. This treatment is used on different types of fibers (glass, etc.) to meet their specific requirements, such as bonding capabilities and the very important protection of fibers. A major requirement for these agents involves the proper handling of the glass fibers during their treatment. Continuous glass fiber (as well as other fiber) strands intended for weaving are usually treated at their forming bushing during their manufacture with starch-oil binders.

Protecting the fibers from damage is achieved through binder lubrication during their formation and during subsequent textile operations, such as twisting, plying, and weaving. Usually they are satisfactory when used with certain TPs but are not compatible with TS plastics. For example, the hydrophilic character of the binders allows moisture to penetrate the glass-plastic interface, which leads to the degradation of RTPs or RTSs in wet and humid environments.

The binder is removed via heat treatment and prior to the application of sizing agents before being used with these plastics. This is accomplished by exposing the reinforcing material (fiber, fabric, etc.) to carefully controlled time-temperature cycles. Chemical sizing coupling agents, such as methacrylic chromic chloride complex and organosilanes, are used to protext the weak heat-cleaned fibers.

A review of the different types of reinforcements is provided in this chapter. A review on the plastics used is included in chapter 2. The popular different forms or compounds of materials used are reviewed in this section.

Prepreg

Prepregs are used to mold different products. The term is used for a reinforcement that is usually preimpregnated with a TS polyester liquid resin. Different forms of reinforcements are used (nonwoven mat, woven fabric, braided, preform, roving, etc.). The catalyzed compounded resin is impregnated into the reinforcement and partially cured to a tack-free state in the B-stage (chapter 1). The reinforcement can be predesigned to meet performance requirements. The molder uses the prepreg in a compression mold or other molding process that will allow the required temperature and pressure conditions to be met, based on how the resin was compounded. With proper storage temperatures (at least about 21°C [70°F]), shelf life can be controlled to be at least 6 months.

Techniques for locating and orienting prepregs onto a molding surface, in accordance with the RP design pattern, are adapted to the tack and drape characteristics of the prepreg. The woven fabrics make the use of sewn stitches, staples, or clamps possible. Usually, the layups are enlarged to provide allowances for trimming after the RP has been cured. Sometimes they are draped over male forms with weighted edges to draw the layups snugly onto the molding surface prior to final cure. Very often, successful layups depend on the operators' skills to innovate.

SHEET MOLDING COMPOUND

Sheet molding compound (SMC) is a ready-to-mold material representing a special form of prepreg. Thicker SMC was developed from the original development of the thin or single-ply prepregs (Table 15.43). It is usually a glass-fiber RTS polyester resin compound in sheet form. The sheet can be rolled into coils during continuous processing. SMC is basically made by mixing and metering the compound, feeding the glass-roving reinforcement, wetting out the glass fibers, rolling up the

Property	SMC chopped fiber	SMC long fiber	Steel SAE 1008
Tensile Strength MPa (psi $\times 10^3$)	65–90 (9.4–13.0)	124–204 (18–29.6)	330.7 (47.9)
Tensile Modulus GPa (psi $\times 10^6$)	10–12.5 (1.2–1.8)	12.2–19.1 (1.8–2.8)	206.7 (30.0)
Flexural Strength MPa (psi $\times 10^3$)	155–200 (22.4–29.0)	248–280 (36.0–55.1)	n/a
Flexural Modulus GPa (psi $\times 10^6$)	8.5–14.0 (1.2–2.0)	11.6–16.4 (1.7–2.4)	n/a
Notched IZOD J/m (ft 1 bs/in.)	500–1000 (9.4–18.7)	725–1360 (13.5–25.4)	n/a
Specific Gravity	1.8–2.0	1.85–2.15	7.86
Coefficient of Expansion m/m.C $\times 10^6$ (in/in.F $\times 10^6$)	12–14 (6.7–7.8)	13–17 (7.2–9.5)	12.1 (6.7)

Table 15.43 Comparing properties of SMC with steel

sheet, and allowing the material to mature. A plastic film separates the layers to enable coiling and to prevent contamination, sticking, and monomer evaporation. This film is removed before the SMC is charged into a mold that may be a matched-die mold or a compression mold.

This moldable material consists primarily of TS polyester resin, glass fiber reinforcement, and filler. Additional ingredients, such as low-profile additives, cure initiators, thickeners, and mold-release agents, are used to enhance the performance or processing of the material. As with any material, such as metallics and plastics, SMC can be formulated in-house or by compounders to meet the performance requirements of a particular application such as tensile properties or Class A surface finishes. Varying the type and percentage of the composition will result in variations in mechanical properties and processibility.

Before SMC can be used for molding, it must age or mature. This maturation time is required to allow the relatively low-viscosity resin to chemically thicken. The thickened SMC is easier to handle and prevents the paste from being squeezed out of the glass fiber sheet during processing. Typically, SMC requires about three to five days to reach the desired molding viscosity.

Different methods are used to produce SMCs that provide different properties and performance (Fig. 15.35). Figure 15.35 depicts an SMC production line using chopped glass fiber for manufacturing thicker SMC (approximately 4 mm thick by 120 cm wide), including roving to provide bidirectional properties, cutting continuous rovings so that it can fit easily in a mold cavity, and a system for producing thicker SMC (about 4 mm thick by 120 cm wide).

Bulk Molding Compound

Bulk molding compound (BMC) is also called dough molding compound (DMC). It is a mixture usually of short glass fibers, resin, and additives similar to the SMC compound. This mixture can be produced in bulk form, or extruded in rope-like form (a log) for easy handling. BMC is available in

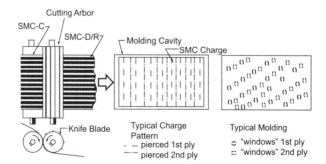

Figure 15.35 These different SMC production lines produce by using chopped glass fibers (top), including roving to provide bidirectional properties, cutting continuous rovings so that they can fit easily in a mold cavity, and producing thicker SMC (about 4 mm thick by 120 cm wide; bottom).

different combinations of resins, usually TS polyesters, additives, and reinforcements. They meet a wide variety of end-use requirements in high-volume applications in which fine finish, good dimensional stability, part complexity, and good overall mechanical properties are important.

BMC is usually molded by compression molding (CM). The RTS BMCs can also be injection molded in much the same way as other plastic compounds using ram, ram-screw, and, for certain BMC mixes, conventional reciprocating screw IM techniques (chapter 4).

Compound

Different RP compounds are used to meet different properties such as directional load. Many TPs (PC, nylon, PP, polystyrene, polyurethane, acetal, polyester TP, polyimide, etc.) are compounded with some type of additive, filler, or plastic blend with reinforcements that are usually short, chopped, or milled glass fibers (383). Commercial RTP compounds are available in several forms: pellets for IM or extrusion, unidirectional tape for FW and similar applications, sheets for stamping and CM, bulk compounds for CM, and so on. Reinforcements significantly improve or modify mechanical properties, whereas fillers are used generally to reduce costs.

Nonfibrous reinforcements are also employed as reinforcements and fillers. They result in increased tensile strength and deflection temperature, but usually decrease impact resistance. Nonfibrous reinforcements are preferred when exceptional flatness is required. Nonfibrous reinforcements include mica, glass beads, and minerals such as wollastonite (talc, calcium carbonate, and kaolin are considered fillers). Unlike fibrous reinforcements, nonfibrous reinforcements can be processed by many different technologies.

There are also flexible RPs. These RTP elastomeric materials provide special engineered products, such as conveyor belts, mechanical belts, high-temperature or chemical-resistant suits, wire and cable insulation, and architectural designed shapes.

RTSs are TS plastics reinforced with fibrous glass, carbon, graphite, or aramid, as well as paper, cotton, flakes, beads, or powders. The workhorse of the RTS industry is TS polyester (also called polyester TS) with glass fiber. The fiber reinforcement may be in the form of chopped fibers; porous, nonwoven mats; woven fabrics; or continuous fibers. The combination of plastics and reinforcements results in versatile materials with unusual characteristics. The reinforcement adds strength and toughness to inherent weather resistance, moldability, and colorability. Thus RTSs are used because of their increased tensile, flexural, torsional, and impact strengths; increased modulus of elasticity; increased creep resistance; reduced coefficient of thermal expansion; increased thermal conductivity; and, in many cases, lower costs.

Other plastics used include epoxy, phenolic, melamine, urea formaldehyde, silicone, and polyurethane TS. They may be formulated to produce a range of materials from soft, flexible elastomers to tough solids. For structural RP applications, the rigid material is of principal interest. They feature outstanding wear and abrasion resistance with low coefficients of friction. Examples of products made with them include truck wheels, wear plates, liners for equipment that handles abrasive materials, and pump impellers.

FABRICATING PROCESS

About 5 wt% of all plastic products produced worldwide are RPs. IM consumes more than 50 wt% of all RP materials with practically all of it being TPs (chapter 4).

They range in fabricating pressures from zero (contact) to moderate to relatively high pressure (2000 to 30,000 psi [13.8 to 207 MPa]), at temperatures based on the plastic's requirements, which may be room temperature or higher. Equipment may be low in cost or rather expensive, such as specialized computer controls of the basic machine together with auxiliary equipment. In turn, labor costs range from very high for low-cost equipment to very low for the high-cost equipment.

Each process offers trade-offs in terms of meeting production quantities (small to large quantities and shapes); performance requirements; the proper ratio of reinforcement to matrix, fiber orientation, reliability, and quality control; surface finish; materials used and quantities of materials used; tolerances; time schedules; and so forth, versus costs (equipment, labor, utilities, etc.). There are products that use only one process, but there can be applications in which different processes can be used.

PREFORM PROCESS

The preform process has been used since the 1940s. As time passed, significant improvements occurred in processing equipment, plastics materials, and cost conditions. This is a method of making chopped-fiber mats of complex shapes that are used as reinforcements in different RP molding and fabricating processes rather than conventional flat mats that may tear, wrinkle, or give uneven glass distribution when producing complex shapes. Most of the reinforcements used are glass fiber rovings. They are desirable where the product to be molded is deep or of very complex shapes. Oriented patterns can be incorporated in the preforms. Different methods are used, with each having many different modifications. Some of these include using a plenum chamber, directed fibers, or a water slurry.

The rovings in a plenum chamber are fed into a cutter. After being cut to the desired lengths, they fall either either into a plenum chamber or onto a perforated screen from under which the air is exhausted. A plastic binder of usually up to 5 wt% is applied and is later cured. As the glass falls into the plenum chamber, the airflow pattern and baffles inside the screen control its distribution. The preform screen rotates and sometimes tilts to ensure maximizing uniform deposits of the roving (Fig. 15.36).

With the directed fiber system, strands are blown onto a rotating preform screen from a flexible hose. The roving is directed into a chopper, where airflow moves it to a preform screen. Use can be made of a vertical or horizontal rotating turntable. This process requires a rather high degree of skill on the part of the operator; however, automated robots are used to provide a controlled system producing quality preforms (Fig. 15.37).

Water slurry is made up of chopped strands in water (takeoff used by the paper pulp industry for centuries). It produces intricately shaped preforms that are tough and self-supporting. Cellulose fibers or bonding resins or both can be used to bond the preform. Where maximum strength is not required, the cellulose content can be high. The fibers can be dyed during the slurry process (Fig. 15.38).

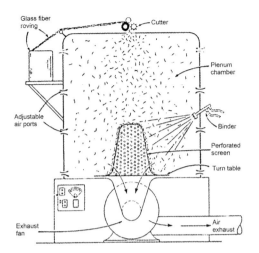

Figure 15.36 Flow of glass fiber rovings traveling through a plenum machine.

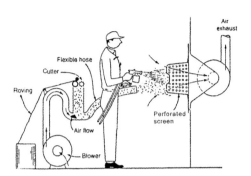

Figure 15.37 Flow of glass fiber rovings traveling through a direct machine.

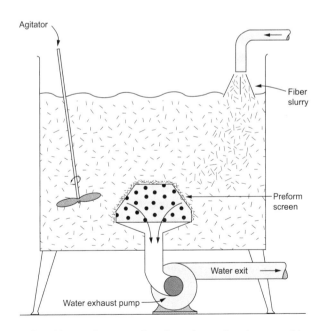

Figure 15.38 Flow of glass fiber rovings traveling through a water-slurry machine.

The screen is important to success. Different shapes can be used to meet different product designs. Cylindrical preforms are easier and less costly to produce than boxlike sections. It is important to recognize that during the rotation of a cylindrical part, the fibrous glass will flow uniformly onto the screen because most sections move at a uniform linear rate. With a rectangular section, this is difficult because the corners rotate in a wider circle than do the center sections and because the airflow is lowest at the corners. Contouring the box shape can improve reinforcement distribution.

Preform screens are usually made from sixteen-gauge perforated material with ⅛ in holes on 3⁄16 in centers. This produces an area that is about 40% open. For some operations, a more open area is required. Perforation patterns are also used to develop specifically designed reinforcement directional properties. The screen is usually designed so that the outside contour is identical with the contour of the mating half of the mold. A screen that is not of the correct size will cause a great deal of difficulty in the molding operation. If the screen is too small, the preform will tear during the molding. If too large, wrinkling and overlapping of the preform will result.

The preform is usually heavy on the flat top and light on the edges and corners. Internal baffles may be added in the preform screen to control the airflow, thus giving a more uniform deposition of glass. The exact area of the baffle usually has to be worked out on a trial-and-error basis until experience is developed. Close cooperation with the preform-machine manufacturer is helpful.

When molding a product with variable wall thickness, it is possible to vary the thickness of the preform. This is usually accomplished by baffling. Another approach is to completely block off areas where no fiber is desired. This action saves material that would otherwise be trimmed off and probably discarded. It is practical to combine two or more preforms into one molded part. This technique is very useful when the thickness of the molded part prohibits the collection of the preform in one piece.

TYPE PROCESS

The more commonly used processes for unreinforced plastics (URPs) also use RPs. They include those reviewed in this book, namely IM, extrusion, thermoforming, foaming, calendering, coating, casting, reaction IM (RIM), rotational molding (RM), CM, transfer molding (TM), and others (chapters 4 to 17). These processes are usually limited to using short reinforcing fibers; however, there are processes that can use long fibers (383). Since glass fibers are extensively used, specifically in IM, the glass fibers will cause wear of metals during processing, such as plasticating barrels and molds or dies. Using appropriate metals that extend their operating time can reduce this wear (chapter 17). The following processes being reviewed are principally designed to process RPs (229 to 231).

COMPRESSION MOLDING

TS plastics in reinforced sheets and compounds are usually used. Also used are RTP sheets and compounds. With TSs, CM can use preheated material (heated with a dielectric heater, for example) that is placed in a heated mold cavity. The mold is closed under pressure, causing the material to flow and fill the cavity. Chemical cross-linking occurs, solidifying the TS molding material (Fig. 15.39).

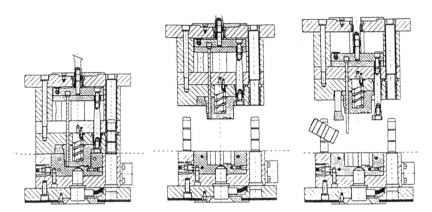

Figure 15.39 Two-part compression mold.

The closed mold shapes the material, usually by heat and pressure. With special additives, the TS material can cure at room temperature. It would have a time limit (pot life) prior to curing and hardening. Based on the compound's preparation, sufficient time is allowed to store and handle the compound prior to its chemical-reaction curing action (chapter 14).

Depending on which plastic is being molded, the clamping force may range from contact to more than thousands of tons. TS polyesters usually have just contact pressure. For plastics requiring pressure, the average range is from 1 to 2 tons. A force that is usually much less than the clamping force is also required to open the mold, so one has to ensure that mold-opening clamping pressure is available. This requirement is usually not a problem. Clamping predominantly uses hydraulic systems. Also popular are all-electric-drive systems and hydraulic-electric hybrid systems. The actual mechanical mechanisms range from toggle to straight-ram systems. Each of these different systems has their individual advantages (chapter 4).

The mold is fastened on the platens. These platens usually include a mold-mounting pattern of bolt holes or T slots; a standard pattern is recommended by the Society of the Plastics Industry (SPI). Platens range from the usual parallel design to other configurations meeting different requirements. The parallel type can include one or more "floating" platens located between the stationary and normal movable platens, resulting in two or more daylight openings where two or more molds or flat laminates can be used simultaneously during one operating cycle. Other designs include shuttles (in which usually two or more molds are moved so that one mold is positioned to receive material and then moves to the press, permitting another mold to receive material and allowing insert molding, shortened molding cycles, etc.), rotary or carousel systems, and "book" opening or tilting presses (7, 12, 183).

With certain plastic compounds, mold breathing is required. This action is also called mold bumping, dwell pause, dwell, gassing, and degassing. It is a pause or a series of repeated pauses in the application of mold pressure using plastics that gives off gases during the heating process; it is employed to remove any entrapped air. This on-off-on pressure action occurs in parts of a second, just prior to the mold closing completely, to allow the escape of gas and air. This stage takes place with many TS plastics, the vulcanization of RTS elastomers, and any material that releases gases.

A CM charging tray, also called a loading tray, is used. The tray is designed to charge simultaneously with material all the mold cavities of a multi-impression mold. The tray has openings where the material is placed (manually or, usually, automatically). In turn, a withdrawing sliding bottom tray initially closes the openings and then slides, exposing openings matching the top tray so that material drops into the cavities.

Applying vacuum in a mold cavity can be very beneficial in molding plastics at low pressures. A press can include a vacuum chamber around or within the mold, providing the removal of air and other gases from the cavities.

FLEXIBLE PLUNGER MOLDING

Flexible plunger molding is a takeoff from CM that uses solid material and male and female matching mold halves (chapter 14). This unique process uses a precision-made, solid-shaped, heated cavity and a flexible plunger that is usually made of hard rubber or polyurethane. This two-part system can be mounted in a press, either hydraulic or air-actuated. Rather excellent product qualities are possible at fairly low production rates. The reinforcement is positioned in the cavity and the liquid TS resin is poured on it. Prepregs, BMC, and SMC are also used in this system.

In flexible plunger molding, a plug is forced into a cavity and the product is cured. The plunger is somewhat deeper and narrower than the cavity. It is tapered in such a manner that contact occurs first in the lowest part of the mold. Ultimate pressures reach up to 400 to 700 kPa (58 to 100 psi), and the plunger causes the contact area to expand radially toward the rim of the cavity, thereby forcing the resin and air ahead of it through the reinforcement. The aim is to develop a void-free product. The pressure conforms to irregularities in the layup, permits wall thickness to be varied within reasonable limits, and makes a good surface possible against a metal mold. The fact that the heat can be applied only from the cavity side leads to long cure cycles, but the same factor tends to produce resin richness, and therefore greater smoothness on the outside of the molding.

FLEXIBLE BAG MOLDING

An air-inflated, pressurized, flexible envelope can replace the plunger. This system provides higher glass content and decreases the chance of voids. Limitations include the need for extensive trimming and that the process creates products with only one good surface.

LAMINATE

Laminates can be many different fabricated RP products, such as high-, contact-, and low-pressure laminates. It usually identifies flat or curved panels using high pressure rather than contact or low pressure. It is a product made by bonding two or more layers of laminate materials. The usual resins are TSs such as epoxies, phenolics, melamines, and TS polyesters. A modification of this process uses TPs. The materials used are limited only by market requirements and include combinations of different woven and/or nonwoven fabrics, aluminum, steel, paper, plastic film, and so on. In the RP industry, laminates refer to fiber-resin materials cured with many different directional properties.

High-pressure laminates generally use preloaded (prepreg) RP sheets in a hot mold at pressures in excess of 7 MPa (1015 psi). Compression, multiplaten presses are used with up to thirty platens to produce flat (also curved) sheets at high production rates. Laminates are molded between each platen simultaneously. Automatic systems can be used to feed material simultaneously between each platen opening, and cured products are automatically removed after curing when the multiple platens open. The contact- or low-pressure laminates use prepregs that cure at low pressures, such as TS polyester resins. Depending on the resin formulation, only contact pressure, such as the pressure applied using hand-operated rollers, is required. Usually, the highest pressure for low-pressure laminates is 350 kPa (50 psi).

These laminates have been used in the industry for almost half a century because of their electrical properties, impact strength, wearing qualities, chemical resistance, their ability to be used as decorative panels, and other important characteristics. (These properties vary depending on the fiber resin used and whether they have a surfacing material.) The major change in the process took place about a half-century ago, when full automation significantly reduced labor costs.

Hand Layup

This low-cost process has different names, such as open molding, contact molding, or bag molding. (Due to different market uses at times different processing names that overlap a process are used). It is a very simple and most versatile process for producing RP products. However, it is slow and is usually very labor intense. It consists of the hand tailoring and the placing of layers of (usually glass fiber) fibrous reinforcements, in either randomly oriented mat, woven roving, or fabric on a one-piece mold and simultaneously saturating the layers with a liquid plastic (usually TS polyester; Fig. 15.40). Usually it is necessary to coat the mold cavity with a parting agent. Gel coatings with or without very thin woven or mat glass fiber scrim reinforcement are also applied to provide smooth and attractive surfaces. Molds can be made of inexpensive metal, plaster, RP, wood, and other materials (chapter 18).

Depending on the resin preparation, the material in or around a mold can be cured with or without heat, and commonly without pressure. Curing needs include room temperature (Fig. 15.6), heat sources, vacuum bags, pressure bags, autoclaves, and so on. An alternative is to use preimpregnated, B stage TS polyester or SMC, but in this case heat is applied with low pressure via an impermeable sheet over the material (Figs. 15.5 and 15.6). This process can produce compact structures that meet tight thickness tolerance, simulating injection molded products.

Generally, the process only requires low-cost equipment that is not automated. However, automated systems have been used. Figure 15.41 shows an automated-integrated hand layup system that uses TS prepreg sheet material. Automation includes cutting and providing the layout of the cut prepreg in a mold. In turn, the designed RP assembly is delivered to a curing station, such as an oven or an autoclave.

This process can be recommended for prototype products, products with small production runs, very large and complex products, and products that require high strength and reliability. The size of the product that can be made is limited by the size of the curing oven. However, sunlight-curing systems (in which UV rays play a role in the curing) and room temperature–curing systems permit practically unlimited product size. Alternate curing methods include induction, infusion, dielectric microwaves, xenon, electron beams, and gamma radiation.

The general process of hand molding can be subdivided into specific molding methods such as those that follow. The terms of some of these methods as well as others reviewed here overlap the same technology; the different terms are derived from different sections of the RP industry.

Vacuum bag molding

This process, also called bag molding, is the conventional hand layup or spray-up that is allowed to cure without the use of external pressure. For many applications this is sufficient, but maximum consolidation may not be reached. There can be some porosity; fibers may not fit closely into internal corners with sharp radiuses but tend to spring back. Resin-rich and resin-starved areas may develop because of draining, even with thixotropic agents. With moderate pressure, these defects or limitations can be overcome with an improvement in mechanical properties.

One way to apply such moderate pressure is to enclose the wet-liquid resin material and mold in a flexible membrane or bag, and draw a vacuum inside the enclosure. Atmospheric pressure on the outside then presses the bag or membrane uniformly against the wet layup. An effective pressure of 69 to 283 kPa (10 to 14 psi) is applied to the product. Air is mechanically worked out of the layup by hand, usually with serrated rollers. The vacuum helps directly by removing air from the wet layup via techniques such as using bleeder channels within the bag (using materials such as jute, glass wool, etc.), which not only removes air but also permits draining of any excess resin. The layup is then exposed to heat using an oven or a heat lamp.

Vacuum bag molding and pressure

To maximize properties in the product, higher pressure is needed in the conventional vacuum bag system. A second envelope can be placed around the whole assemblage. Air under pressure is admitted between the inner bag and the outer envelope after the initial vacuum cycle is completed. Still higher uniform pressures can be obtained by placing the vacuum assemblage in an autoclave. An initial vacuum may or may not be employed in this technique. The result derived from using an autoclave is encouraging.

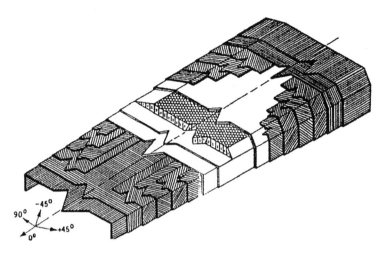

Figure 15.40 Layout of reinforcement is designed to meet structural requirements.

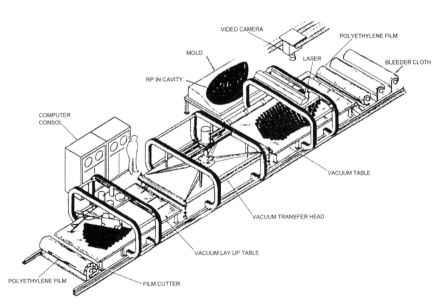

Figure 15.41 Automated-integrated RP vacuum hand layup process that uses prepreg sheets that are in the B-stage (chapter 1).

Pressure bag molding

This process is used when more pressure is required than in those processes just reviewed. A second envelope (or structure) is placed around the whole assemblage and air pressure admitted between the inner bag and outer envelope, or between the inner bag and the structure. The application of pressure (using air, steam, or water) forces the bag against the product to apply pressure while the product cures. Using this combination of vacuum and pressure bags assists in air or gas removal and higher pressures, thus resulting in more densification.

Autoclave molding

Very high pressures can be obtained for processing RPs by placing a pressure bag or vacuum bag molding assemblage in an autoclave. This curing process may or may not employ an initial vacuum. Some of the different RP processes are used in conjunction with an autoclave oven (Fig. 15.42) Hot-air or steam pressures of 0.36 to 1380 MPa (50 to 200 psi) are used. The higher pressure yields denser products. If still higher pressures are required (avoid this approach unless you have considered the danger of extremely high pressures), a hydroclave may be used, employing water pressures as high as 70 MPa (10150 psi). The bag must be well sealed to prevent infiltration of high-pressure air, steam, or water into the molded product. In all of these approaches, the fluid pressure adjusts to irregularities in the layup and remains effective during all phases of the resin cure, even though the resin may shrink. This process can be used to make seamless contains, tanks, and pipes.

Autoclave press

This process simulates autoclaving by using the platens of a press to seal the ends of an open chamber. It provides both the force required to prevent loss of the pressurized medium and the heat required to cure the RP inside.

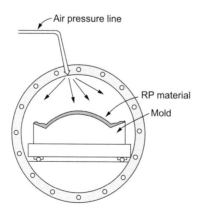

Figure 15.42 Schematic of hand-layup bag molding in an autoclave.

Wet layup

This method is sometimes combined with bag molding to enhance the properties; this procedure is usually just called bag molding. Because it is difficult to wet out dry fibers with too little resin, initial volumetric fraction ratios of resin to fiber are seldom less than two to one. On a weight basis, the ratio is about one to one. Liquid catalyzed resin is handworked or automatically worked into the fibers to ensure the wet-out of fibers and to reduce or eliminate entrapped air.

Bag molding hinterspritzen

This patented process allows virgin or recycled TPs, such as PP and PC/ABS, to thermally bond with the backing of multilayer PP-based fabrics; this process provides good elasticity. This one-step molding technique provides a low-cost approach for in-mold fabric laminations that range from simple to complex shapes.

Contact molding

Contact molding is also called open molding or contact pressure molding. It is a process for molding RPs in which the reinforcement and plastic are placed in a mold cavity. Depending on the plastic used, curing takes place at either room temperature, using a catalyst-promoter system, or by heating in an oven without pressure or using very little (contact) pressure. Contact molding gave rise to bag molding, hand layup or open molding, and low-pressure molding. It plays a significant role in molding RPs. It is difficult to surpass if a few products are to be made at the lowest cost. The process is basically what was reviewed in the section that covered bag molding.

Filament Winding

The use of circumferential wrappings to increase the bursting strength of certain structures is not new. Historically, wire and tape (Fig. 15.43) wrappings have been used to prevent the bursting of cannon barrels and to wrap wooden pipes both to increase the bursting strength and to hold the two parts together so that a leakproof cylinder forms. The use of filamentary structures for applications requiring ultimate structural performance is rather recent and unique (Tables 15.44 and 15.45).

FW is a fabrication technique for forming RP parts of high strength and light weight. It is made possible by exploiting the remarkable strength properties of their continuous fibers or filaments encased in a matrix of a resinous material. For this process, the reinforcement consists of filamentous nonmetallic or metallic materials processed either in fibrous or tape forms.

Some form of glass, including continuous filaments roving, yarn, or tape, is used. The glass filaments, in whatever form, are encased in a plastic matrix, either wetted out immediately before winding (wet process) or impregnated ahead of time (preimpregnated process). The plastic contains the reinforcement, holds it in place, seals it from mechanical damage, and protects it from environmental deterioration. The reinforcement–matrix combination is wound continuously on a form or

Railway tank cars	Irrigation pipes
Storage tanks: acids, alkalies, water, oil, salts, etc.	Salt water disposal pipes
	Underground water pipe
High-voltage switch gears	Oil well tubes
Electrical containers	Ladders
Propellers	Extension arms for telephone trucks
High-pressure bottles	Textile bobbins
Decorative building supports	Weather rockets
Containers for engines, batteries, etc.	Gas bottle-mines
Buoys	Structural tubing
Valves	Insulating tubes
Aircraft tanks	Electrical conduit
Aircraft under-carriage	Chemical pipe
Aircraft structures	Pulp and paper mill pipe
Fishing rods	Water heating tanks
Round nose boat	Pipe fittings and elbows
Boat masts	Truck-mounted booms
Lamp poles	Highway stanchions
Golf clubs	Capacitor jackets and spacers
Race track railing	Coil forms
Auto bodies	Electronic waveguides
Drive shafts	Printed circuit forms
Air brake cylinder	Electric motor rotors, binding bands
Heating ducts	Circuit breaker housing
Acid filters	High-voltage insulators
Recoil-less rifle barrel	Rectifier spacers
Pontoons	Antenna/dishes
Motor housing	Rotating armatures–DC motors
Computer housings	DC commutator
Marker buoys	Fan housing
Laundry tubs	High voltage fuse tubes
Ventilator housings	Floating ducts
Rifle barrel	Automotive parts
Dairy cases	Tank trucks
Auto and truck springs	Light poles
Circuit breaker rupture pots	Brassiere supports
Cartop boats	Looms
Electroplating jigs	

Table 15.44 Filament-wound structures for commercial and industrial applications

mandrel whose shape corresponds to the inner structure of the part being fabricated. After the matrix cures, the form may be discarded or it may be used as an integral part of the structural item.

Reinforcements have set pattern layups to meet performance requirements (Table 15.46 and Figs. 15.44 and 15.45). The aim is to have them uniformly stressed. Figure 15.27 shows the relationship of RP density to the percentage of glass fiber by weight or volume, which can be related to the compacting action that occurs when FW takes place.

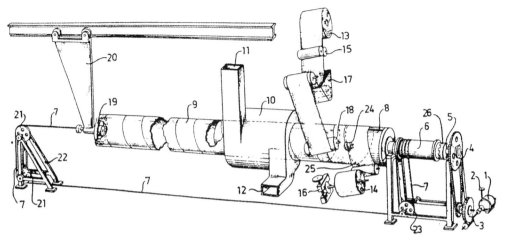

Figure 15.43 Early-twentieth-century tape-wrapping patent of a tube-making machine by Hoganas-Billesholms A.B., Sweden.

Rocket motor cases	Liquid rocket thrust chamber
Rocket motor insulators	Rocket exit cones
Solid propellent motor liners	Chemical rockets
Nose cones for space fairings	Chemical tanks
Rocket nose cones (2)	Sounding rocket tubes
Rocket nozzle liners	Tactical bombardment rockets
Jato motor	Tent poles
APU turbine cases	Heat shields
High-pressure bottles (gas or liquid)	Artillery shell shipping grommet
Vacuum cylinders	Artillery round-protective cones
Torpedo launching tubes	Submarine fluid pipes
Rocket launcher tubes	Submarine tanks and containers
Flame thrower tubes	Submarine ventilation pipes
Missile landing spikes	Submarine hulls
Deep space satellite structures	Underwater buoys
Radomes	Cryogenic vessels
Igniter baskets	Electronic packages
Wing dip tanks	Submarine fairwaters
Helicopter rotor blades	Sonar domes
Thermistors	Engine cowlings
Missile shipping cylinders	Fuse cases
Boat ventilator cowlings	Torpedo cases and launchers

Table 15.45 Filament-wound structures for aerospace, hydrospace, and military applications

Type of winding	Considerations	Machinery required
Hoop or circumferential	High winding angle. Complete coverage of mandrel each pass of carriage. Reversal of carriage can be made at any time without affecting pattern.	Simple equipment. Even a lathe will suffice.
Helix with wide ribbon	Complete coverage of mandrel each pass of carriage. Reversal of carriage can be made at any time without affecting pattern.	Simple equipment with provision for wide selection of accurate ratios of carriage-to-mandrel speeds. Powerful machine and many spools of fiber required for large mandrel.
Helix with narrow ribbon and medium or high angle	Multiple passes of carriage necessary to cover mandrel. Programmed relationship between carriage motion and mandrel rotation necessary. Reversal of carriage must be timed precisely with mandrel rotation. Dwell at each end of carriage stroke may be necessary to correctly position fibers and prevent slippage.	Precise helical winding machine required. Ratio of carriage motion to mandrel rotation must be adjustable in very small increments. Relationship of carriage to mandrel positions must be held in selected program without error through carriage reversals and dwells. Relationship between carriage position and mandrel rotation must be progressive so that pattern will progress.
Helix with low winding angle	Fibers positioned around end of mandrel close to support shaft. Characteristics of "helix with narrow ribbon" apply. Fibers tend to go slack and loop on reversal of carriage. Fibers tend to group from ribbon into rope during carriage reversal. Mandrel turns so slowly that extremely long delay occurs at each end of carriage stroke and speed-up of mandrel at each end of carriage stroke is highly desirable to shorten winding time.	Similar machinery required as for "helix with narrow ribbon." Take-up device for slack fibers is necessary if cross-feed on carriage is not used. Cross-feed on carriage is required for very low winding angles. Programmed rotating eye can be used to keep ribbon in flat band at carriage reversal. Mandrel speed-up device must be programmed exactly with carriage motion or pattern will be lost. Polar wrap machine can be used for narrow ribbons with winding angle below about 15° without take-up device or mandrel speed-up being required.

Table 15.46 Different FW patterns meet different performance requirements

Type of winding	Considerations	Machinery required
Zero or longitudinal	Mandrel must remain motionless during pass of carriage and then rotate a precise amount near 180° while carriage dwells. Fibers must be held close to support shaft during mandrel motion or fibers will slip.	Precise mandrel indexing required. Simple two-position cross feed on carriage sufficient. Vertical mandrel machine and pressure follower for ribbon sometimes required to preserve ribbon integrity.
Polar wrap	Low angle wrap. Fibers may be placed at different distances from centers at each end when geodesic (non-slipping) path does not have to be followed.	Polar wrap machine with swinging fiber delivery arm desirable for high-speed winding. Helical machine with programmed cross-feed will wind polar wraps more slowly.
Cone	General considerations same as for helical winding except that carriage motion is not uniform.	Programmed non-linear carriage motion required. Other machine requirements same as for helical winding.
Simple spherical	Planar windings at a particular angle result in a heavy build-up of fibers at ends of wrap. For more uniform strength, successive windings at higher angles are required.	Sine wave motion of carriage is required for carriage with no cross-feed. At low angles of wind, cross-feed is necessary because carriage travel becomes excessive. Polar wrap machine may be used if range of axis inclination is large enough.

Table 15.46 Different FW patterns meet different performance requirements *(continued)*

Type of winding	Considerations	Machinery required
Simple ovaloid	Similar to simple spherical winding but with different carriage or cross-feed motion.	Helical machine with programmed carriage or cross-feed. Polar wrap machine can be used where geodesic (non-slipping) path is in a plane.
True spherical	Path of fibers programmed to give uniform wall thickness and strength to all areas on sphere.	Special machine best approach. Otherwise complex programming of all motions of helical machine required.
Miscellaneous	For successful filament winding, it must be possible to hand-wind with no sideways slipping of fibers on mandels surface.	Machine to reproduce motions of hand winding. Programmed motions in several axes may be required.

Table 15.46 Different FW patterns meet different performance requirements *(continued)*

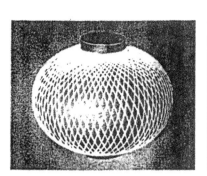

Figure 15.44 Views of fiber filament-wound isotensoid pattern of the reinforcing fibers without plastic (left) and with resin cured.

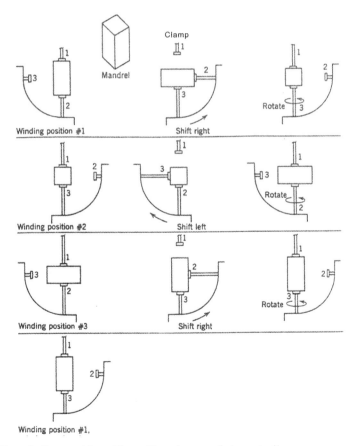

Figure 15.45 Box winding machine with position changes of clamp tooling.

In winding cylindrical pressure vessels, tanks, or rocket motors, two winding angles are generally used (Figs. 15.3 and 15.8). One angle is determined by the problem of winding the dome integrally with the cylinder. Its magnitude is a function of the geometry of the dome. These windings also pick up the longitudinal stresses. The other windings are circumferential or 90° to the axes of the case and provide hoop strength for the cylindrical section.

It is comparatively easy to wind domes with a single polar port integrally with a cylinder and without the necessity of cutting filaments. Cutting is obviously not desirable, since it interrupts the continuity of the orthotropic material. The usual procedure in winding multiported domes is to add interlaminate reinforcements during the winding operation, where the ports are to be located.

It is possible to wind integrally most of the bodies of revolution, such as spheres, oblate spheres, and torroids. Each application, however, requires a study to insure that the winding geometry satisfies the membrane forces induced by the configuration being wound.

FW can be carried out on specially designed automatic machines. Precise control of the winding pattern and direction of the filaments are required for maximum strength, which can be achieved only with controlled machine operation. The equipment in use permits the fabrication of parts in accordance with properly designed parameters so that the reinforced filamentous wetting system is in complete balance and optimal strength is obtained. The maximum strength is achieved when filaments in tension carry all major stresses. Under proper design and controlled fabrication, hoop tensile strengths in filament-wound items of over 3500 MPa (508,000 psi) can be achieved, although a strength of 1500 MPa (218,000 psi) is more frequently achieved.

Since this fabrication technique allows the production of strong, lightweight parts, it has proved particularly useful for components of structures of commercial and industrial usefulness (Table 15.44) and for aerospace, hydrospace, and military applications (Table 15.45). Both the reinforcement and the matrix can be tailor-made to satisfy almost any property demand. This aids in widening the applicability of FW to the production of almost any item wherein the strength-to-weight ratio is important. FW is used in creating products of different shapes, including circles, ellipses, and rectangles.

FW structures present certain problems because of the lack of ductility in the glass reinforcement. These problems can be partially solved by proper design and fabrication procedures. Reinforcements other than glass can be used to obtain good ductility, but some of these have lower temperature strength and characteristics. Proper construction constitutes a well-proved means of utilizing an intrinsically nonductile reinforcement to obtain a high degree of confidence in the structural integrity of the end product. Since glass has high strength and is a relatively low-cost product, glass filaments are still the major reinforcing materials. Other filaments for applications requiring properties such as higher temperatures or greater stiffness include quartz, carbon, graphite, ceramics, and metals alone or in combinations that include glass fibers.

A further difficulty with the basic materials is that they do not lend themselves readily to simple concepts and to simple comparisons. The matrix components are essentially the same plastics as those used for conventional RP laminates. Epoxy plastics are more widely used than others, although phenolics and silicones give structures with higher temperature properties. TS polyesters are used for many commercial structures in which cost is a problem and high temperatures are not needed.

For certain FW vessels, the glass-plastic material's low modulus of elasticity is a serious disadvantage. Only moderate improvements in modulus of elasticity, achieved through modifications in glass composition or in processing, tend to be feasible. Any significant improvement in the modulus of elasticity requires changes in the glass composition. There are effective additives to the glass, such as beryllium oxide, that increase its modulus without causing a proportional increase in density.

Interlaminar shear constitutes possible limitations on FW parts. Although the absence of interweaving (fabrics) boosts tensile strength by eliminating cross-fraying, shear strength is limited by the bonding of the reinforcement to the plastic. In conventional woven cloth laminates, the high points of one layer tend to interlock with the low points of adjacent layers. This results in strengthening the composite against shear failure. Compared to other plastics or matrices, epoxy gives better interlaminar shear because of its inherently better bonding. By proper design, the values of interlaminar shear can be minimized.

FW structures have lower ultimate bearing strengths than conventional laminates, for they are more rigid and less ductile. Accordingly, they have less ability to absorb stress concentrations around holes and "cut-outs." The original higher tensile strength permits allowable design stresses under these conditions. Since cutting, drilling, or grooving for attachments or access openings reduces the high mechanical strength of filament-wound structures, proper design is necessary. Damaging machine operations are to be avoided after final curing of the part. Destructive cut-outs or attachment holes are to be eliminated by incorporating the use of premolded plastic or metal inserts into the designs.

Techniques cannot be used for every structural element. The shape of the part must permit the removal of the winding mandrel after final curing. Reversed curvatures should be eliminated whenever possible, since it is difficult to wind them and hold the filaments under tension. In order to meet this problem, fusible, expandable, and multiparty mandrels are often required.

The cost of FW parts is low only when volume production is achievable. Manufacturing processes should be mechanized and completely automated to obtain, by extensive and careful tooling, the close tolerances that are required in filament-wound structures to meet high-strength and low-cost objectives (Fig. 15.46). Precision winders with carefully selected mandrels and speed controls, special curing ovens, and matched grinders are required. It takes time to develop this equipment, and a high initial investment is necessary. Once the original tooling cost has been amortized, the unit cost of individual filament-wound parts becomes relatively low, since the basic materials have a low cost.

Fabricating RP tank

Classical stress analysis proves that hoop stress (stress trying to push out the ends of the tank) is twice that of longitudinal stress. The construction of a tank made with conventional materials (steel, aluminum, etc.) requires the designer to use sufficient materials to resist the hoop stresses that result in unused strength in the longitudinal direction. In RP, however, the designer specifies a laminate that has twice as many fibers in the hoop direction as in the longitudinal direction.

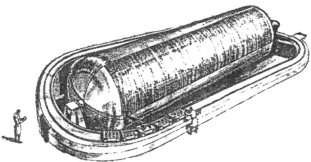

Figure 15.46 Schematics of "racetrack" filament-winding machines. Top view shows machine in action; other view is a schematic of a machine built to fabricate 150,000 gal rocket motor tanks.

Consider a tank 0.9 m (3 ft) in diameter and 1.8 m (6 ft) long with semispherical ends. Such a tank's stress calculations (excluding the weight of both the products contained in it and the support for the tank) are represented by the following formulas:

$$s = pd/2t \text{ for the hoop stress} \qquad (15\text{-}1)$$

and

$$s = pd/4t \text{ for the end and longitudinal stresses,} \qquad (15\text{-}2)$$

where s = stress, p = pressure, d = diameter, and t = thickness.

Tensile stresses are critical in tank design. The designer assumes the pressure in this application will not exceed 100 psi (700 Pa) and selects a safety factor of 5. The stress must be known so that the thickness can be determined. The stress or the strength of the final laminate is derived from the makeup and proportions of the resin, mat, and continuous fibers in the RP composite material.

Representative panels must be made and tested, with the developed tensile stress values then used in the formula. Thus the calculated tank thickness and method of layup or construction can be determined based on

$$t_h = pd/2s_h \text{ or } t_h = pd/4s_h, \qquad (15\text{-}3)$$

where

$$t_h = \frac{100 \times 3 \times 12}{2 \times \dfrac{20 \times 10^3}{5}} = 0.450 \text{ in},$$

$t_l = 1/2 \, t_h = 0.225$ in (or the same; thickness with half the load or stress),

$$t_h = \text{hoop thickness},$$

$$t_l = \text{longitudinal thickness},$$

$$s_h = \text{hoop stress},$$

$$s_l = \text{longitudinal stress}, \qquad (15\text{-}4)$$

$$s_h = 20 \times 10^3 \text{ psi (140 MPa)},$$

$$\text{safety factor} = 5,$$

$$p = 100 \text{ psi (700 Pa)},$$

$$d = 3 \text{ ft (0.9 m)}.$$

If the stress values had been developed from a laminate of alternating plies of woven roving and mat, the layup plan would include sufficient plies to make 1 cm (0.40 in) or about four plies of woven roving and three plies of 460 g/m² (1½ oz) mat. However, the laminate would be too strong axially. To achieve a laminate with a hoop-to-axial strength of 2 to 1, one would have to carefully specify the fibers in those two right-angle directions, or filament wind the tank so that the vector sum of the helical wraps would give a value of 2 (hoop) and 1 (axial), or a wrap of approximately 54° from the axial.

Another alternative would be to select a special fabric whose weave is 2 to 1, wrap to fill, and circumferentially wrap the cylindrical sections to the proper thickness, thus getting the required hoop and axial strengths with no extra, unnecessary strength in the axial (longitudinal) direction, as would inevitably be the case with a homogeneous metal tank.

The design of RP products, while essentially similar to conventional design, does differ in that the materials are combined when the product is manufactured. The RP designer must consider how the load-bearing fibers are placed and ensure that they stay in the proper position during fabrication.

INJECTION MOLDING

As reviewed in chapter 4, over 50 wt% of all RPs go through conventional IM machines (IMMs; Fig. 15.47). These use practically all TPs (chapter 4). The RP compounds that are thick and pasty (BMC, for example) are principally processed through ram IMMs with some going through screw IMMs (Figs. 15 48 and 15.49).

As explained in chapter 4, both short and long glass and other fibers are injection molded. Most of the long fiber RPs utilizes the two-stage machine design that includes replacing a preloader with a single- or twin-screw extruder.

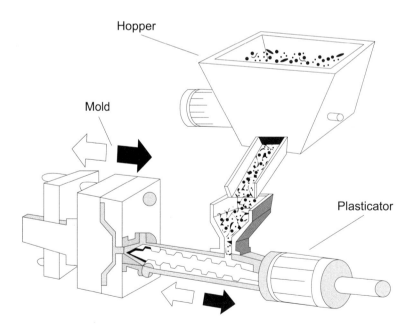

Figure 15.47 Conventional single stage IMM.

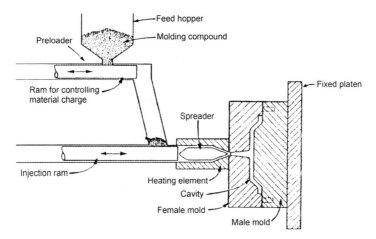

Figure 15.48 IM with a preloader usually providing heat to the RP compound.

MARCO PROCESS

Between the 1940s and 1960s, this process was used to fabricate many different RP products. It was the takeoff for RTM and bag molding. Reinforcements are laid up in any desired pattern, as in RTM and bag molding. Low-cost matched molds (wood, for example) confine the reinforcement. In this process, the usual liquid catalyzed TS polyester surrounds the mold in its open trough (Fig. 15.50). From a central opening (hole) in one of the mold halves, a pressure is applied so that the plastic flows through the reinforcements. With the proper wet-out of fibers, voids are eliminated.

This method, when first used, was the reverse of RTM. By 1960 the Marco method used vacuum pressure at the parting line and also used a vacuum push-pull action, where pressure was applied in the center hole similar to what is now used in RTM. Pressure was applied through the center hole alone or in combination with a vacuum from the trough area to aid the flow of the liquid plastic.

PULTRUSION

This process can produce products that meet very high structural requirements, high weight-to-strength performances, electrical requirements, and so on (389). It is a continuous process for fabricating RPs that usually has a constant cross-sectional shape (I-, U-, H-, and other shapes). The reinforcing fibers are pulled through a plastic (usually TS) liquid impregnation bath through rollers and then through a shaping die followed by a curing action. The material most commonly used is TS polyester with glass fiber. Other plastics, such as epoxy and polyurethane, are used when their improved properties are needed. When required, fiber material in mat or woven form is added for cross-ply properties.

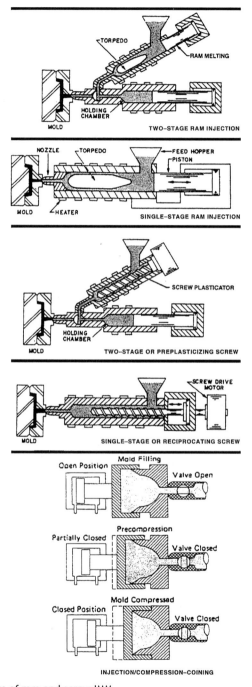

Figure 15.49 Schematics of ram and screw IMM.

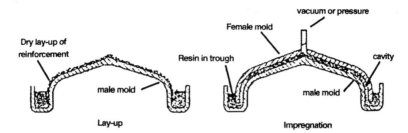

Figure 15.50 Use is made of vacuum, pressure, or pressure-vacuum in the Marco process.

There are also systems that eliminate the plastic bath, in which the plastic is impregnated in the die. This approach is a variation on extrusion-wire and cable-coating systems that provide controlled impregnation (chapter 5). Cleverly designed dies that include rotating sections to provide complex pultruded products have been used.

In contrast to extrusion, this process involves a combination of liquid resin and continuous fibers (or a combination with short fibers) that is pulled continuously through a heated die of the shape required for continuous profiles. Glass content typically ranges from 25 wt% to 75 wt% for sheet and shapes, and at least 75 wt% for rods. RP shapes include I-beams, L-channels, tubes, angles, rods, sheets, and so on.

Reactive Liquid Molding

Reactive liquid composite molding (RLCM) proceeds in two steps: (1) preform formation by organizing loose fibers into a shaped preform and (2) impregnation of the fibers with a low-viscosity reacting liquid. Heat transfer in the mold may thermally activate the reacting material or mixing activated by impingement of two reactive streams as in the polymerization of polymers (chapter 1). Simulations of flow and reaction, a relatively recent innovation in RLCM, allow determination of vent and weld line locations, fill times, and control of racetracking in terms of gate locations (when IM takes place), mat permeability, and processing conditions. Commercial success requires (1) fast reaction and (2) efficient preform formation. Using higher mold temperatures and preheating the preform can decrease cycle time for thermally active systems (178).

Innovative processes for preforming include the following:

1. A thermoformable mat, heated by IR to melt the binder, is pressed into shape by one or two moving platens while supported by a hold/slip edge clamp to reduce wrinkling.
2. Automated, directed fiber performers employ multiple delivery systems to create a surface veil, a chopped roving layer, and continuous roving with loops, all of which are fused by hot air.

3. In the Seeman Composites Resin Infusion Process (SCRIMP), channels of resin flow between layers of fibers or along internal networks.
4. With water slurry deposition, fibers are positions by water flow through a contoured screen that sets them with hot air.

Innovations to reduce costs by combining process steps include the following:

1. Parts are formed directly from sheets (i.e., by heating a porous sheet and then consolidating and shaping it in a compression mold).
2. The hot air preformer produces preforms by either directed fiber or TP mat forming.
3. The cut-and-shoot process combines preforming and molding steps consecutively in the same tool.
4. The bladder inflation inside a mold shapes preforms and forms the mold wall during filling.

Low pressure and temperature processing by RLCM allow the use of inexpensive, lightweight tools, especially for prototyping. RLCM allows customizing reinforcement to give desired local properties and part consolidation via complex 3-D geometries.

Its applications include marine and poolside products, sanitary ware, caskets, automotive panels, and vehicle suspension links. Materials used include isocyanate-based resins (mixing activated), unsaturated polyester, and styrene (thermal activated).

Reinforced RTM

Reinforced RIM (RRIM) is a closed-mold, low-pressure process in which a preplaced dry reinforcement fiber construction (such as woven and nonwoven fabric or a fiber preform) with or without decorative surface material is impregnated with a liquid plastic (usually TS polyurethane; chapter 12) through an opening in the center area of a mold (Fig. 15.51). The resin, at about 50 psi (0.3 MPa) pressure, moves through the reinforcement located in the mold cavity. The air inside the cavity is displaced by the advancing resin front and escapes through vents located at the high points or the last areas of the mold to be filled. When the mold has filled, the vents and the resin inlets are closed. After curing via room-temperature hardeners or heat or both, the part is removed. This process provides a rather simple approach to molding designed RP parts in relatively low-cost molds (using low pressure), and the molds are manufactured in a short time. It can also incorporate vacuum to assist resin flow; with vacuum-assisted RTM, the process is called infusion molding. This process could be identified as a takeoff of the Marco process.

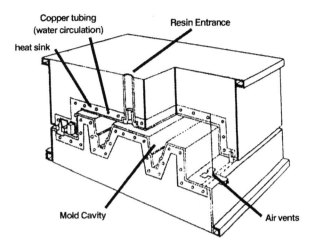

Figure 15.51 Cutaway view of a reinforced RTM mold.

Reinforced Rotational Molding

In reinforced rotational molding (RRM), the solid (powder or pellet) or liquid, with or without reinforcing fibers, and TP materials are used. is the material is placed in a mold that only has a cavity to form the outside of the part to be made. The mold is rotated simultaneously about two axes at similar or different speeds, depending on the part configuration. The material is forced against the walls of the cavity. It first goes through a heating period to melt the resin, followed by a cooling period to solidify the plastic. Small to large parts are molded. Because molds are not subjected to pressure, relatively low-cost molds can be used (471).

Squeeze Molding

This method is a takeoff between RTM and the hand layup process. The reinforcement and a room-temperature-curing TS polyester resin are put into a mold. In turn, the mold is put into an air-pressure bag, where the resin is slowly forced through the reinforcement in the mold cavity at low pressures of about 200 to 500 kPa (30 to 75 psi). The RP is cured at room temperature in unheated molds. It is a slow process that can mold only a few products per day.

SCRIMP Process

SCRIMP is a gas-assisted RTM process. Glass fiber fabrics/TS vinyl ester polyester plastic and polyurethane foam panels (for insulation) are usually used. They are placed in a segmented tool. A vacuum is pulled with a bag so that a huge amount of plastic can be drawn into the mold. It is similar to various RP molding processes. It is adaptable to fabricating large RP products, such as buses weighing about 10000 kg (22000 lb), about 3200 kg (7000 lb) lighter than steel units.

SOLUBLE CORE MOLDING

This technology is also called fusible core technology, soluble core technology (SCT), lost-wax molding, loss-core molding. This technique is a takeoff of and is similar to the lost-wax molding process used by the ancient Egyptians to fabricate jewelry. In this process, a core is usually molded with a eutectic alloy with a low melting point (e.g., zinc and tin), water-soluble TP, wax, and so on. During core installation, it can be supported by the mold core pins, spiders, or other tools. The core is inserted into a mold (using IM, CM, casting, etc.) and plastic is injected or located around the core. When the plastic has solidified and is removed from the mold, the core is removed through an existing opening by melting at a temperature below the plastic's melting point. If there is no existing opening, a hole has to be drilled.

LOST-WAX PROCESS

When this soluble fusible core molding technique was first used, it involved a bar of wax wrapped with RPs (such as glass fiber-TS polyester resin; Fig. 15.52). After the RP is cured (via bag molding, an oven, an autoclave, etc.) in a restricted mold to keep the rectangular shape, the wax is removed at low heat by drilling a hole or removing the ends. The result is a very high-strength RP channel. Its shape can be rectangular, round, curved, and so on. It was used in 1944 to fabricate the first all-plastic airplane. This lost-wax process was used with bag molding the RP sandwich monocoque construction.

SPRAY-UP

This process has been popular with RP production for over half a century. With time passing, significant new developments took place, particularly in the spraying equipment. An air-spray gun

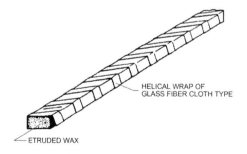

Figure 15.52 Lost-wax process fabricated a high-strength RP structural beam.

includes a roller cutter that chops glass fiber rovings to a controlled short length before the fiber is blown in a random pattern onto a mold surface. This action can be manual or automatic. Suppliers of spray-up equipment continue to produce equipment with cleaner, reduced-styrene emissions (as low as 2.2%), higher capacity, more uniform spray patterns, and more versatility (390). There are many types of spray guns, including external or internal mixing guns, distributive/turbulent mixing guns, air-atomized guns, airless guns, and others.

As the fibers leave the spray gun simultaneously, the gun sprays the usual catalyzed TS polyester plastic (with styrene monomer; chapter 2). The chopped fibers are coated in plastic as they exit the gun's nozzle (Fig. 15.53). The resulting fluffy RP mass is consolidated with serrated rollers to squeeze out air and reduce or eliminate voids. A closed mold with appropriate temperature and pressure produce the products. Figure 15.53 shows only a few types of guns: (1) a type 3 with a three-layer approach, (2) a type 2 airless bullet type, and (3) a type 2 airless glass type.

Figure 15.53 Nonatomized, dispensed Glass-Craft spray gun is easy to use and produces low styrene emissions and is economic to maintain.

STAMPING

RTS plastic B stage sheet material can be used with its required heating cycle (chapter 1). However, the most popular is the RTP sheets usually using PP plastics (chapter 19). Compared to IM RTPs, these stamped products can provide improved mechanical and physical properties, such as impact strength, heat-distortion temperature, and much less anisotropy.

The RP sheet material is precut to the required size, which depends on the size of the part to be molded. The precut sheet is preheated in an oven; the heat required depends on the TP used (such as PP or nylon, where the heat can range from 270°C to 315°C [520°F to 600°F]). Dielectric heat is usually used to ensure that the heat is quick and, most important, to provide uniform heating through the thickness and across the sheet. After heating, the sheet is quickly formed into the desired shape in cooler matched-metal dies, using conventional metal-stamping presses or SMC-type compression presses.

Stamping is potentially a highly productive process capable of forming complex shapes with the retention of the fiber orientation in particular locations as required. The process can be adapted to a wide variety of configurations, from small components to large box-shaped housings and from flat panels to thick, heavily ribbed parts.

COLD FORMING

This process is similar to the hot-forming stamping process. It is a process of changing the shape of a plastic sheet or billet in the solid phase through plastic (permanent) deformation with the use of pressure dies. The deformation usually occurs with the material at room temperature. However, it also includes forming at a higher temperature or warm forming, but much below the plastic melt temperature, and lower than the temperatures used in thermoforming or hot stamping.

Different forms of glass fiber-TS plastics are used with or without special surface coatings such as gels. Materials are compounded with controlled pot life so that they start their cure reaction after being placed in the mold cavity. Room-temperature curing occurs by an exothermic chemical reaction that heats the RP. Pressures are moderate at about 140 to 350 kPa (20 to 50 psi). Molds can be made of inexpensive metal, plaster, RP, wood, and other materials.

COMOFORM COLD MOLDING

This is another version of cold forming that utilizes a thermoformed plastic skin to impart an excellent surface and other characteristics (weather resistance, for example) to a cold-molded RP. For example, a TP sheet is placed in a matched-mold cavity with an RP uncured material placed against the sheet. The mold is closed and the fast, room-temperature-curing resin system hardens. The finished product has a smooth TP-formed sheet backed up with RP.

SELECTING PROCESSES

The different processes available for fabricating RPs each tend to have their own specific performance capabilities and costs. It is important to recognize that the processes can have significant effects on the performances of the finished product. Because more than one process exists, the choice may involve studying the repeated output quality of each process. The choice can be related to the type of plastic to be processed, pressure and temperature requirements in the curing process, quantity of products needed, the size of product needed, production rates, tolerances required, and so on. Each process, like each material of construction, has capabilities and limits. Tables 15.47 to 15.52 provide information on different processes with properties and characteristics of RPs.

Process problems that may need to be addressed include insufficient compaction and consolidation before plastic solidification, cure occurring before air pockets develop, incomplete or uncontrollable wet-out and encapsulation of the fibers, and/or insufficient fiber or nonuniform fiber content. These deficiencies lead to loss of strength and stiffness and susceptibility to deterioration by water and aggressive agents. Heat control may not be adequate, particularly for crystalline plastics, or it may be too rapid (chapter 1).

In some applications, the designer or fabricator will not have the ability to choose freely from all the design, material, and process alternatives. For example, a design is often heavily constrained by the need to fit an existing assembly, and the material and process may be determined largely by the need to use existing fabrication facilities.

The geometric symmetry of a product can influence process selection. Both shape and design details are closely related to the process choice. The ability to mold ribs, for example, may depend on material flow during a process or on the flowability of a plastic reinforced with glass. The ability to produce hollow shapes depends on the ability to use removable cores of air, fusible or soluble solids, and even sand. Hollow shapes can also be produced using cores that remain in the product, such as foam inserts in RTM or metal inserts in IM.

A process's pressure and the available equipment can limit product size, whereas the ability to achieve specific shape and design detail is dependent on the way the process operates. Generally, the lower the processing pressure, the larger the product that can be produced. With most labor-intensive methods, such as hand layup, slow-reacting TSs can be used, and there is virtually no limit on size.

There may be a requirement for surface finish, molded-in color, textured surface, or other conditions that the plastic material is to meet (chapter 2). Surface finish can be an important consideration. The different processes may be able to provide only one surface to be smooth or both sides to be smooth. It is important that smoothness be identified because it has many meanings to different people. Surface finish can be more than just a cosmetic standard. It also affects product quality, mold cost, and delivery time. Standards set by the Society of Plastics Engineers (SPE) and the Society of the Plastics Industry (SPI) range from a No. 1 mirror finish to a No. 6 grit blast finish. A mold finish comparison kit consisting of six hardened tool steel pieces and associated molded pieces is available through the SPE and SPI.

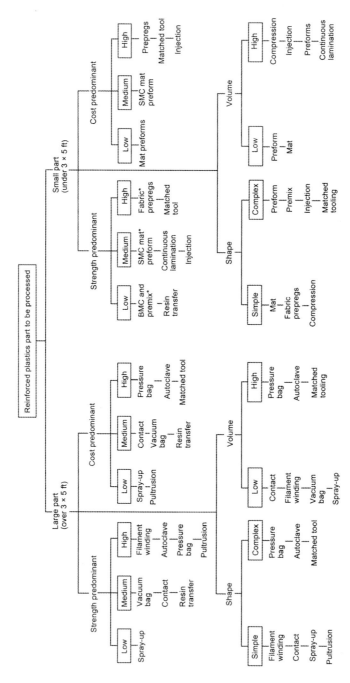

Table 15.47 RP processing guide to RP process selection

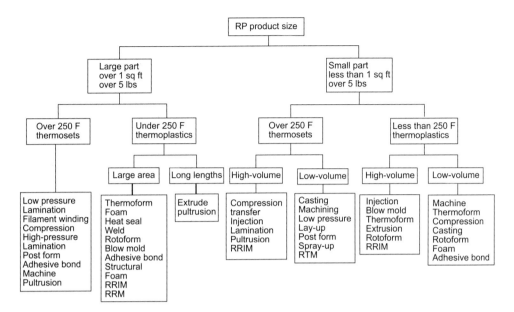

Table 15.48 RP processing guide to RP size

DESIGN

In order to understand potential problems and solutions of design, it is helpful to consider the relationships of machine capabilities, plastics processing variables, and part requirements (Tables 15.53 to 15.59 and Figs. 15.54 to 15.56). A distinction needs to be made here between machine conditions and processing variables. For example, machine conditions include the operating temperature and pressure, mold temperature, machine output rate, and so on. Processing variables are more specific, such as the material melt condition in the mold, the flow rate versus the temperature, and so on.

Aspect Ratio

The aspect ratio is the ratio of length to diameter (L/D) or length to thickness (L/T) of a material such as a fiber; it is also the ratio of the major to minor axis lengths of a material such as a particle. These ratios can be used to determine the effect of dispersed additive fibers or particles on the viscosity of a fluid or a melt and in turn on the performance of the compound based on L/D ratios. In RPs, fiber L/D will have a direct influence on the RP's performance.

As shown in Figure 15.14, L/Ds with high values of 5 to 10 provide for high strength in RPs. Lining up and overlapping fibers and disks take advantage of directional properties. Theoretically, with proper layup, the most high-performance plastics could be obtained when compared to other materials.

Process	Description	Comments
Hand layup	Polyester and epoxy; glass cloth	Simplest method, minimum equipment, inexpensive mold, short run production items; boats, aircraft, prototypes.
Spray-up	Polyesters and epoxies; chopped glass roving	Boats, swimming pools, tank lining.
Bag	Glass mat (plus binder resin) shaped to contour of final mold	Economical production method—lends self to complex shapes.
Compression	Storage stable reinforcement preimpregnated with resin, B-staged or partially advanced; phenolics, diallyl phthalates, polyesters, epoxies	Uniform and easy to handle—eliminates shop formulation; molded at point and time of application by standard molding processes; electrical, sporting goods, aircraft, aerospace.
Filament winding	Glass roving or strands, impregnated with resin, are wound continuously and uniformly on mandrel; polyesters, epoxies	High strength-to-weight ratio; tanks, pipe, aerospace.
Vacuum bag	Uses atmospheric pressure, layup covered with cellophane or polyvinyl alcohol and vacuum dawn	Short precision run products
Pressure bag	Variation of vacuum bag wherein pressure and heat are applied to the layup during cure	Improved resin distribution and surface finish relative to vacuum bag
Autoclave	Variation of pressure bag wherein entire layup is placed in steam autoclave and pressure raised up to 700 psi	Can handle higher glass-to-resin ratios; uniform product
RRIM, RRM, etc.	Resin and reinforcement molded under heat and pressure to produce parts with excellent uniformity and surface finish: polyurethane, polyesters, epoxies, diallyl phthalates	Fast, automated production line procedure; boats, chairs, trays, car fenders and side panels, etc.

Table 15.49 Examples of a few processes to material comparisons

	Process		
	RTM	SMC compression	Injection
Process operation:			
Production requirement, annual units per press	5,000–10,000	50,000	50,000
Capital investment	Moderate	High	High
Labor cost	High	Moderate	Moderate
Skill requirements	Considerable	Very low	Lowest
Finishing	Trim flash, etc.	Very little	Very little
Product:			
Complexity	Very complex	Moderate	Greatest
Size	Very large parts	Big flat parts	Moderate
Tolerance	Good	Very good	Very good
Surface appearance	Gel coated	Very good	Very good
Voids/wrinkles	Occasional	Rarely	Least
Reproducibility	Skill dependent	Very good	Excellent
Cores/inserts	Possible	Very difficult	Possible
Material usage:			
Raw material, cost	Lowest	Highest	High
Handling/applying	Skill dependent	Easy	Automatic
Waste	Up to 3 percent	Very low	Sprues, runners
Scrap	Skill dependent	Cuts reusable	Low
Reinforcement flexibility	Yes	No	No
Mold:			
Initial cost	Moderate	Very high	Very high
Cycle life	3,000–4,000 parts	Years	Years
Preparation	In factory	Special mold-making shops	
Maintenance	In factory	Special machine shops	

Table 15.50 RP resin transfer, SMC compression, and IM processes compared

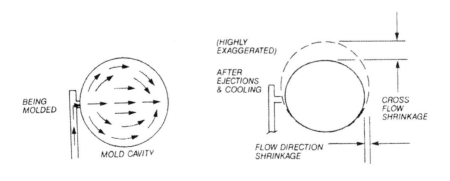

Figure 15.54 Example of the effect of shrinkage in the longitudinal and transverse directions of a molded part.

Thermosets	Properties	Processes
Polyesters	Simplest, most versatile, economical and most widely used family of resins, having good mechanical properties, electrical properties, good chemical resistance, especially to acids	Compression molding Filament winding Hand layup Mat molding Pressure bag molding Continuous pultrusion Injection molding Spray-up Centrifugal casting Cold molding Encapsulation
Expoxies	Excellent mechanical properties, dimensional stability, chemical resistance (especially alkalis), low water absorption, self-extinguishing (when halogenated), low shrinkage, good abrasion resistance, very good adhesion properties	Compression molding Filament winding Reaction injection molding Hand layup Continuous pultrusion Encapsulation
Phenolics	Good acid resistance, good electrical properties (except arc resistance), high heat resistance	Centrifugal casting Compression molding
Silicones	Highest heat resistance, low water absorption, excellent dielectric properties, high arc resistance	Continuous laminating Compression molding Injection molding
Melamines	Good heat resistance, high impact strength	Encapsulation Compression molding
Diallyl phthalate	Good electrical insulation, low water absorption	Compression molding

Table 15.51 Examples of RTS plastic processes

Material family	Injection	Compression	Transfer	Casting	Cold molding	Coating	Structural foam	Extrusion	Laminating	Sheet forming	RP molding FRP	Filament	Dip and slush	Blow	Rotational
ABS	X						X	X		X				X	
Acetal	X							X		X				X	X
Acrylic	X	X						X		X				X	
Allyl			X	X	X				X	X	X	X			
ASA	X						X	X	X	X					
Cellulosic	X							X		X					X
Epoxy		X	X	X	X	X			X	X	X	X			
Fluoroplastic	X	X	X		X	X		X	X						X
Melamine-formaldehyde	X	X	X	X		X			X						
Nylon	X			X		X	X	X						X	X
Phenol-formaldehyde	X	X	X	X	X	X			X						
Poly(amide-imide)	X	X	X												
Polyarylether	X							X							
Polybutadiene	X		X					X							
Polycarbonate	X	X					X	X		X			X	X	X
Polyester (TP)	X					X		X						X	X
Polyester-fiberglass (TS)		X	X					X			X	X			
Polyethylene	X	X				X	X	X	X	X				X	X
Polyimide	X	X				X		X(TP)	X						X
Polyphenylene oxide	X						X	X							X
Polyphenylene sulfide	X	X					X								
Polypropylene	X						X	X	X	X				X	X
Polystyrene	X						X	X		X				X	X
Polysulfone	X	X						X		X					
Polyurethane (TS)	X		X	X		X	X	X(TP)	X					X	X
SAN	X						X	X	X	X					
Silicone		X		X		X		X	X						X
Styrene butadiene	X					X		X	X					X	X
Urea formaldehyde		X						X	X						
Vinyl	X					X	X	X	X	X			X	X	X

* Compounding permits using other processes.

Table 15.52 Comparing uses of different plastics with different RP and other processes

Design parameter	Resin-transfer molding	Spray-up	Hand lay-up	Sheet molding compound
Minimum inside radius, in. (mm)	$\frac{1}{4}$ (6.35)	$\frac{1}{4}$ (6.35)	$\frac{1}{4}$ (6.35)	$\frac{1}{16}$ (1.59)
Molded-in holes	No	Large	Large	Yes
In-mold trimming	No	No	No	Yes
Core pull and slides	Difficult	Difficult	Difficult	Yes
Undercuts	Difficult	Difficult	Difficult	Yes
Minimum recommended draft (deg.)	2 to 3	0	0	1 to 3; 3, or as
Minimum practical thickness, in. (mm)	0.080 (2.0)	0.060 (1.5)	0.060 (1.5)	0.050 (1.3)
Maximum practical thickness, in. (mm)	0.500 (12.7)	No limit	No limit	1 (25.4)
Normal thickness variation, in. (mm)	±0.010 (±0.25)	±0.020 (±0.50)	±0.020 (±0.50)	±0.005 (±0.1)
Maximum thickness buildup, heavy buildup (ratio)	2:1	Any	Any	Any
Corrugated sections	Yes	Yes	Yes	Yes
Metal inserts	Yes	Yes	Yes	Yes
Bosses	Difficult	Yes	Yes	Yes
Ribs	Difficult	No	No	Yes
Hat section	Yes	Yes	Yes	No
Raised numbers	Yes	Yes	Yes	Yes
Finished surfaces	2	1	1	2

Table 15.53 Examples of interrelating product-RP material-process performances

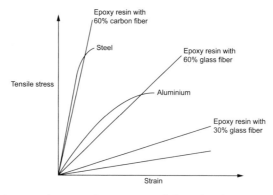

Figure 15.55 Tensile stress-strain curves for epoxy-unreinforced and epoxy-reinforced RPs and other materials.

	Resin transfer molding	Injection molding	Pultrusion	Reinforced reaction-injection molding
factor limiting maximum size of product	machine size	machine size	materials	metering equipment
maximum size, m²	9.3	9.3	29	4.6
shape limitations	moldable	moldable	round, rectangular	moldable
production cycle time	10–20 min	15 s to 15 min	10–30 min	1–2 min
glass, %	15–25	20–40	30–75	5–25
strength orientation	random	random	highly oriented	with flow
strength	low–medium	low	high	low
wall thickness, mm				
minimum	0.76	0.76	1.6	2.0
maximum	25	13–25	13	13
tolerance			0.3 ± 25	± 0.05
variations	uniform	uniform	uniform	uniform
minimum draft				
to 15 cm depth	1°	1°	0–2°	1–3°
over 15 cm depth	1°	1° +	0–2°	3° +
minimum inside radius	½ part depth	½ part depth	1.5 mm	½ part depth
ribs	yes	yes	no	yes
bosses	yes	yes	no	yes
undercuts	possible	possible	no	yes
holes				
parallel	yes	yes	no	yes
perpendicular	yes	yes	no	yes
built-in cores	yes	yes	no	yes
metal inserts	yes	yes	no	yes
metal edge stiffeners	yes	yes	no	yes
surface finish				
number of finished surfaces	all	2	2	2
quality of surface	excellent	excellent	fair to good	excellent
gel-coat surface		yes	no	no
surfacing mat	yes	yes	no	no
combination with thermoplastic liner	yes	yes	no	no
trim in mold	no	no	no	yes
molded-in labels	yes	yes	no	no
raised numbers	yes	yes	no	yes
translucency	yes	yes	no	no
tool cost	high	high	low	low–medium
capital equipment cost	high	high	low	low–medium

Table 15.54 Comparison of RP design aspects and processes to cost

Hand lay-up, spraying	Filament winding	SMC	BMC
mold size; transport	winding machine	press rating and size	press rating and size
280	93	4.6	4.6
none	surface of revolution	moldable	moldable
3 min to 24 h	5 min	1.5–5 min	1.5–5 min
20–35	65–90	15–35	15–35
random	highly oriented	random	random
medium	very high	low–medium	low–medium
0.76	0.3	0.8	1.5
≥38	50	6.4	25
+0.5	+0.3	+0.2	+0.1
as desired	as desired	uniform desirable <3:1	as desired
0–2°	3°	1–3°	1–3°
0–2°	3° +	3° +	3° +
6.3 mm	3.1 mm	1.5 mm	1.5 mm
yes	no	yes	yes
yes	no	yes	yes
avoid	no	avoid	no
yes	yes	yes	yes
yes		undesirable	undesirable
possible	possible	possible	possible
yes	yes	yes	yes
yes	no	no	no
1	1	2	all
excellent	excellent	very good	excellent
yes	yes	no	no
yes	yes	no	no
yes	yes	no	no
no	yes	yes	yes
yes	yes	difficult	difficult
yes		yes	yes
yes	yes	no	no
low	low	high	high
low	low	high	high

Table 15.54 Comparison of RP design aspects and processes to cost *(continued)*

Molding		Materials	Flexural modulus × 10³ kg/mm³	Impact strength	Heat resistance (HDT 18.6)°C	Paintability by baking (140-150°C)	Weight ratio[1] (Equiflextural modulus)	Moldability	Cost[2] (Mold cost)[3]
Compression molding	Hot press	Polyester + GF (SMC)	1	○~△	>200	◎	0.65	○~△	○ (○)
		Polyester + GF (BMC)	1.1	↑	↑	◎	0.6	↑	↑
		Polyester + GF (High-strength SMC)	1.6~4.2	◎	↑	—	0.4~0.5	↑	○ (○)
	Cold press	Polyester + GF (Resin injection)	0.8	○	150~200	△~×	0.62	△	◎ (◎)
		Polyester + GF (Hand lay-up)	—	—	—	—	—	—	—
	Stamping	PP + GF or sawdust, paper pulp (AZDEL etc.)	0.6	◎	160	—	0.5	◎	△~○ (○)
		Nylon + GFTF (STX, etc.)	0.8	◎	215	○	↑	◎	△ (○)
Filament winding		Epoxy + CF (CFRP)	15	○	>200	—	0.2	△	△~× (○)
Injection molding		PP + GF, talc (EPDM) AS + GF	0.6~0.4	↑	120~105	—	0.5	○	◎ (◎)
		PBT or nylon + GF	1.2~1.4	↑	205~215	◎~○	0.5	↑	(—)
		Foamed styrene or ABS (+ GF)	2.4~2.5	○~△	80 (100)	—	0.4~0.6	↑	△ (○) ○ (○)
RLM		Urethane + GF (RRIM)	0.1~0.2	◎	—	△~×	—	○~△	△ (◎~○)

Note: 1. Ratio based on sheet metal weight as 1; 2. Relative comparison for 400-500 kg; 3. Mold cost for sheet metal. Symbols: ◎ Excellent; ○ Good; △ Fair; × No good

Table 15.55 Examples of processing variables

Part design	Casting	Compression	Filament winding	Injection	Matched die molding	Rotational	Transfer compression	Wet lay-up (contact molding)
Major shape characteristics	Simple configurations	Moldable in one plane	Structure with surfaces of revolution	Few limitations	Moldable in one plane	Hollow bodies	Simple configurations	Moldable in one plane
Limiting size factor	Material	Equipment	Equipment	Equipment	Equipment	Material	Equipment	Mold size
Maximum thickness, in. (mm)	None	0.5 (12.7)	3 (76)	6 (150)	2 (51)	0.5 (12.7)	6 (150)	0.5 (12.7)
Minimum inside radius, in. (mm)	0.01–0.125 (0.25–3.18)	0.125 (3.18)	0.125 (3.18)	0.01–0.125 (0.25–3.18)	0.06 (1.5)	0.01–0.125 (0.25–3.18)	0.01–0.125 (0.25–3.18)	0.25 (6.4)
Minimum draft (deg.)	0–1	>1	2–3	<1	1	1	1	0
Minimum thickness, in. (mm)	0.01–0.125 (0.25–3.18)	0.01–0.125 (0.25–3.18)	0.015 (0.38)	0.005 (0.1)	0.03 (0.8)	0.02 (0.5)	0.01–0.125 (0.25–3.18)	0.06 (1.5)
Threads	Yes	Yes	No	Yes	No	Yes	Yes	No
Undercuts	Yes[1]	NR[2]	NR[2]	Yes[1]	NR[2]	Yes[1]	NR[2]	Yes
Inserts	Yes	Yes	Yes	Yes	Yes	Yes	Yes	Yes
Built-in cores	Yes	No	Yes	Yes	Yes	Yes	Yes	Yes
Molded-in holes	Yes	Yes	No	Yes	Yes	Yes	Yes	Yes
Bosses	Yes	Yes	No[5]	Yes	No[6]	Yes	Yes	Yes
Fins or ribs	Yes	Yes	No	Yes	Yes	Yes	Yes	Yes
Molded-in designs and nos.	Yes	Yes	No	Yes	Yes	Yes	Yes	Yes
Surface finish[7]	2	1–2	5	1	4–5	2–3	1–2	4–5
Overall dimensional tolerance (in./in., plus or minus)	0.001	0.001	0.005	0.001	0.005	0.01	0.001	0.02

[1]Special mold required. [2]Not recommended. [3]Only flexible material. [4]Only direction of extrusion. [5]Possible with special techniques. [6]Fusing premix/yes. [7]Rated 1 to 5 (1 = very smooth, 5 = rough).

Table 15.56 Product design versus processing methods

	Contact molding, spray-up	Pressure bag	Filament winding	Continuous pultrusion	Matched die molding with preform or mat
Minimum inside radius, in.	$\frac{1}{4}$	$\frac{1}{2}$	$\frac{1}{8}$	$\frac{1}{16}$	$\frac{1}{8}$
Molded-in holes	Large	Large	Yes	Yes	Yes
Trimmed-in mold	No	No	Yes	No	Yes
Built-in cores	Yes	Yes	Yes	Yes	Yes
Undercuts	Yes	Yes	No	No	No
Minimum practical thickness, in. (mm)	0.060 (1.5)	0.060 (1.5)	0.010 (0.25)	0.037 (0.94)	0.030 (0.76)
Maximum practical thickness, in. (mm)	0.50 (13)	1 (25.4)	3 (76.2)	1 (25.4)	0.25 (6.4)
Normal thickness variation, in. (mm)	±0.020 (±0.51)	±0.020 (±0.51)	±0.010 (±0.25)	±0.005 (±0.1)	±0.008 (±0.02)
Maximum buildup of thickness	As desired	As desired	As desired	Yes	2 to 1 maximum
Corrugated sections	Yes	Yes	Circumferential only	In longitudinal direction	Yes
Metal inserts	Yes	Yes	Yes	No	Yes
Surfacing mat	Yes	Yes	Yes	Yes	Yes
Limiting size factor	Mold size	Bag size	Lathe bed length and swing	Pull capacity	Press dimensions
Metal edge stiffeners	Yes	N.R.	Yes	No	Yes
Bosses	Yes	N.R.	No	No	Yes
Fins	Yes	Yes	No	Yes	N.R.
Molded-in labels	Yes	Yes	Yes	Yes	Yes
Raised numbers	Yes	Yes	No	No	Yes
Gel coat surface	Yes	Yes	Yes	No	Yes
Shape limitations	None	Flexibility of the bag	Surface of revolution	Constant cross-section	Moldable
Translucency	Yes	Yes	Yes	Yes	Yes
Finished surfaces	One	One	One	Two	Two
Strength orientation	Random	As desired	Orientation of ply	Depends on wind	Directional
Typical glass percent by weight	Random 30–45	45–60	50–75	30–60	30

Table 15.57 Other product design considerations versus processing methods

	Compression molding		Cold press molding	Spray-up and hand lay-up	Resin transfer molding
	Sheet molding compound	Pre-form molding			
Minimum inside radius, in.	1/16	1/8	1/4	1/4	1/4
Molded-in holes	yes	yes	no	large	no
Trimmed in mold	yes	yes	yes	yes	yes
Core pull & slides	yes	no	no	no	no
Undercuts	yes	yes	no	yes	yes
Minimum recommended draft, in./deg	1/4 in. to 6 in. depth: 1 deg to 3 deg; 6 in. and over: 3 deg or as required		2 deg 3 deg	0 deg	2 deg 3 deg
Minimum practical thickness, in.	0.050	0.030	0.030	0.060	0.080
Maximum practical thickness, in.	1	0.250	0.500	no limit	0.500
Normal thickness variation, in.	±0.005	±0.008	±0.010	±0.020	±0.010
Maximum thickness build-up, heavy build-up & increased cycle	as req'd.	2-to-1 max.	2-to-1 max.	as req'd.	2-to-1 max.
Corrugated sections	yes	yes	yes	yes	yes
Metal inserts	yes	not recommended	no	yes	yes
Bosses	yes	yes	not recommended	yes	yes
Ribs	as req'd.	not recommended	not recommended	yes	yes
Molded-in labels	yes	yes	yes	yes	yes
Raised numbers	yes	yes	yes	yes	yes
Finished surfaces (reproduces mold surface)	two	two	two	one	two

Table 15.58 Product design shapes versus processing methods

Composite	Tensile strength, psi × 10³ (MPa)	Flexural strength, psi × 10³ (MPa)	Flexural modulus, psi × 10⁶ (GPa)
Hand lay-up/spray-up	10.0 (68.9)	17.0 (117.2)	1.0 (6.9)
Bulk molding compound	3.0 (20.7)	10.0 (68.9)	1.6 (11.0)
Sheet molding compound	10.0 (68.9)	22.0 (151.7)	1.5 (10.3)
Mat/preform-structural	15.0 (103.4)	30.0 (206.8)	1.3 (9.0)
Mat/perform-low shrink	10.5 (72.4)	24.0 (165.5)	1.1 (7.6)
High-glass sheet molding compounds*	19.5 (134.4)	36.0 (248.2)	1.9 (13.1)
XMC-3 composite (lengthwise)	75 (517)	125 (862)	5.5 (37.9)
Pultrusion (lengthwise)	30.0 (206.8)	30.0 (206.8)	2.5 (17.2)

* Minimum averages for a range of high-strength compounds

Table 15.59 Examples of the efficiency RPs fiber orientation

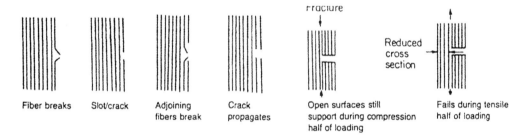

Figure 15.56 Example of crack propagation to fracture that can occur, resulting in product failure under load.

TOLERANCE

The TS plastics and RTSs are very suitable to tight tolerance requirements. Amorphous and crystalline TPs (chapter 1), RTPs, and particularly unreinforced TPs (UTPs) can be more complicated tolerancewise if the fabricator does not understand their behavior. Crystalline plastics generally have different rates of shrinkage in the longitudinal (melt flow direction) and transverse directions, as shown in Figure 15.54, where it is injection molded.

Shrinkage changes can occur at different rates in different directions. These directional shrinkages can vary significantly due to changes in processes such as during IM (chapter 4). Activity is influenced by factors such as injection pressure, melt heat, mold heat, part thickness, and part shape. The amorphous type of melt flow can be easier to balance.

Shrinkage is caused by a volumetric change in a material, particularly RTP, as it cools from a molten to a solid form. Shrinkage is not a single event since it can occur over a period of time for certain plastics, particularly TPs. Most of it happens in the mold, but it can continue for up to 24 to

48 hours after molding. This so-called postmold shrinkage is when a product might be constrained in a cooling fixture. Additional shrinkage can occur, most often with RTPs, when annealing or exposure to high service temperatures relieves frozen-in stress.

The main considerations in mold design in terms of product shrinkage, for instance, with IM of RTPs, are to provide adequate cooling, proper gate size and location, and structural rigidity. Of these three, cooling conditions are the most critical, especially for crystalline TPs.

Certain plastics, such as TS polyester during cross-linking (curing), generate heat that is controlled by constituents such as styrene content; this, in turn, influences shrinkage (chapter 1). In small batches, heat is generated in a controlled manner. In larger batches, the heat generated can cause discoloration or cracking. Modifiers can be added to lower the cross-linking rate. The formation of the cross-linked network is accompanied by some volume contraction (Table 15.60). TSs with high styrene content crack as a more rigid structure attempts to shrink. Fillers, inorganic extenders, and fibers reduce the shrinkage and can eliminate internal voids and cracking (chapter 2).

A number of the computer-aided flow simulation programs offer modules designed to forecast product shrinkage (and, to a limited degree, warpage) from the interplay of plastic and mold temperatures, cavity pressures, molded-part stress, and other variables in mold-fill analysis. The predicted shrinkage values in various areas of the product should be used as the basis for sizing the mold cavity, either by manual input or with a feedthrough to a mold-dimensioning program. All the programs can successfully predict a certain amount of shrinkage.

To meet tolerances or shrinkages (as with other materials), more than simple arithmetic needs to be applied. An important requirement is that someone, usually the manufacturer of the mold, be familiar with plastics behavior and, particularly, the fabrication method to be used. Of course, with experience on a similar product, as with other materials, setting tolerances and shrinkages is almost automatic. Examples of tolerances are given in Tables 15.26, 15.61, and 15.62.

TSs have a much lower shrinkage that is easier to control than that of TPs. RPs have much less shrinkage compared to URPs. Thus with RPs, tolerances and shrinkages are significantly reduced or eliminated, and they provide more reliability in ease of repeatability than URPs, particularly with RTSs.

TS matrices with all types of reinforcements and fillers generally are more suitable to meet tight dimensional tolerances. For example, injection-molded products can be held to extremely close tolerances of less than a thousandth of an inch (0.0025 cm) or down to zero (0.0%). Tolerances that

Monomer	Shrinkage, %
Styrene	17
Vinyltoluene	12.6
Diallyl phthalate	11.8
Methyl methacrylate	21
Vinyl acetate	27

Table 15.60 Example of TS polyester volume shrinkage during curing

Molding method	Thickness range Min. mm (in)	Thickness range Max. mm (in)	Maximum practicable buildup within individual part	Normal thickness tolerance. mm (in)
Hand lay-up	1.5 (0.060)	30 (1.2)	No limit; use cores	±0.5 (0.020)
Spray-up	1.5 (0.060)	13 (0.5)	No limit; use many cores	±0.5 (0.020)
Vacuum-bag molding	1.5 (0.060)	6.3 (0.25)	No limit; over three cores possible	±0.25 (0.010)
BMC molding	1.5 (0.060)	25 (1)	Min. to max. possible	±0.13 (0.005)
Matched-die molding: SMC	1.5 (0.060)	25 (1)	Min. to max. possible	±0.13 (0.005)
Pressure-bag molding	3 ($\frac{1}{8}$)	6.3 ($\frac{1}{4}$)	2 : 1 variation possible	±0.25 (0.010)
Centrifugal casting	2.5 (0.100)	$4\frac{1}{2}$% of diameter	5% of diameter	±0.4 mm (0.015 in)
Filament winding	1.5 (0.060)	25 (1)	Pipe, none; tanks, 3 : 1 around port	Pipe, ±5%; tanks, ±1.5 mm (0.060 in)
Pultrusion	1.5 (0.060)	40 (1.6)	None	1.5 mm, ±0.025 mm ($\frac{1}{16}$ in. ±0.001 in); 40 mm. ± 0.5 mm ($1\frac{1}{2}$ in, ±0.020 in)
Continuous laminating	0.5 (0.020)	6.3 ($1\frac{1}{4}$)	None	±10% by weight
Injection molding	0.9 (0.035)	13 (0.5)	Min. to max. possible	±0.13 (0.005)
Rotational molding	1.3 (0.050)	13 (0.5)	2 : 1 variation possible	±5%
Cold stamping	1.5 (0.060)	6.3–13 (0.25–0.50)	3 : 1 possible as required	±6.5% by weight; ±60% for flat parts

Table 15.61 RPs wall-thickness tolerances

can be met commercially go from 5% for 0.020 in (0.05 cm) thick, to 1% for 0.500 in (1.27 cm), to ½% for 1.000 in (2.54 cm), to ¼% for 5.000 in (12.70 cm), and so on.

Tolerances should not be specified to be tighter than necessary for economical production. However, after production starts, the aim is to mold as tight as possible to be more profitable; using less material and shortening the molding cycle time result in lower fabrication costs. There are unreinforced molded plastics that change dimensions (shrink) immediately, in a day, or in a month; these differences in times are due to material relaxation and changes in temperature, humidity, or load application or all three. RPs can significantly reduce or even eliminate this postmolding dimensional change.

Using any calculated shrinkage approach provides a guide in simple shapes. For other shapes, some critical key dimensions of the product will, more often than not, not be as predictable from the shrink allowance, particularly if the product is long or complex (or both), or has tight tolerances. This situation also exists with other materials (steel, aluminum, etc.). Determining shrinkage involves more than just applying the appropriate correction factor from a material's data sheet. Data sheets provide guides.

	Avg. rate* per ASTM D 955	
Material	0.125 in. (3.18 mm)	0.250 in. (6.35 mm)
ABS		
Unreinforced	0.004	0.007
30% glass fiber	0.001	0.0015
Acetal, copolymer		
Unreinforced	0.017	0.021
30% glass fiber	0.003	NA
HDPE, homo		
Unreinforced	0.015	0.030
30% glass fiber	0.003	0.004
Nylon 6		
Unreinforced	0.013	0.016
30% glass fiber	0.0035	0.0045
Nylon 6/6		
Unreinforced	0.016	0.022
15% glass fiber + 25% mineral	0.006	0.008
15% glass fiber + 25% beads	0.006	0.008
30% glass fiber	0.005	0.0055
PBT polyester		
Unreinforced	0.012	0.018
30% glass fiber	0.003	0.0045
Polycarbonate		
Unreinforced	0.005	0.007
10% glass fiber	0.003	0.004
30% glass fiber	0.001	0.002
Polyether sulfone		
Unreinforced	0.006	0.007
30% glass fiber	0.002	0.003
Polyether-etherketone		
Unreinforced	0.011	0.013
30% glass fiber	0.002	0.003
Polyetherimide		
Unreinforced	0.005	0.007
30% glass fiber	0.002	0.004
Polyphenylene oxide/PS alloy		
Unreinforced	0.005	0.008
30% glass fiber	0.001	0.002
Polyphenylene sulfide		
Unreinforced	0.011	0.004
40% glass fiber	0.002	NA
Polypropylene, homo		
Unreinforced	0.015	0.025
30% glass fiber	0.0035	0.004
Polystyrene		
Unreinforced	0.004	0.006
30% glass fiber	0.005	0.001

Table 15.62 Comparing unreinforced and RP mold shrinkage rates

ENGINEERING ANALYSIS

Design Theory

Fibrous RPs differ from many other engineering materials because they combine two essentially different materials, fibers and synthetic resin, into a single composite. In this they are somewhat analogous to reinforced concrete, which combines concrete and steel, but in RPs the fibers are generally much more evenly distributed throughout the mass and the ratio of fibers to resin is much higher than the ratio of steel to concrete. In the design of fibrous, RPs, it is necessary to take into account the combined action of fiber and resin. Sometimes the combination can be considered to be homogeneous and, therefore, to be similar to engineering materials like metal. In other cases, homogeneity cannot be assumed and it is necessary to take into account the fact that two widely dissimilar materials have been combined into a single unit (chapter 19).

In designing fibrous RPs, certain important assumptions are made. The first and most fundamental is that the two materials act together and that the stretching, compression, and twisting of fibers and of resin under load is the same—that is, the strains in fiber and resin are equal. This assumption implies that a good bond exists between resin and fiber to prevent slippage between them and to prevent wrinkling of the fiber.

The second major assumption is that the material is elastic (that is, the strains are directly proportional to the stresses applied), and when a load is removed the deformation disappears. In engineering terms, the material is assumed to obey Hooke's law. This assumption is probably a close approximation of the actual behavior in direct stress below the proportional limit, particularly in tension, if the fibers are stiff and elastic in the sense of Hooke's law and carry essentially all the stress. The assumption is probably less valid in shear, where the resin carries a substantial portion of the stress. The resin may undergo plastic flow leading to creep or to the relaxation of stress, especially when stresses are high.

More or less implicit in the theory of materials of this type is the assumption that all the fibers are straight and unstressed or that the initial stresses in the individual fibers are essentially equal. In practice it is quite unlikely that this is true. It is to be expected, therefore, that as the load is increased, some fibers reach their breaking points first, and as they fail, their loads are transferred to other unbroken fibers, with the consequence that failure is caused by the successive breaking of fibers rather than by the simultaneous breaking of all of them. The effect is to reduce the overall strength and to reduce the allowable working stresses accordingly, but the design theory is otherwise largely unaffected as long as essentially elastic behavior occurs. The development of higher working stresses is, therefore, largely a question of devising fabrication techniques to make the fibers work together to obtain maximum strength.

Design theory shows that the values of a number of elastic constants must be known in addition to the strength properties of the resin, the fibers, and their combination. Reasonable assumptions are made in carrying out designs. In the examples used, more or less arbitrary values of elastic constants and strength values have been chosen to illustrate the theory. Any other values could be used just as well. See Tables 15.63 and 15.64.

Fiber orientation	Fiber length	A_{fT}	A_m	F_{Theor}* psi	F_{Tests}† psi	Composite efficiency ‡
Filament-wound fibers	Continuous	0.77	0.23	310,000	180,000	58.0%
Cross-laminated fibers	Continuous	0.48	0.52	197,000	72,500	36.8%
Cloth-laminated fibers	Continuous	0.48	0.52	197,000	43,000	21.8%
Mat-laminated fibers	Continuous	0.48	0.52	197,000	57,200	29.0%
Chopped-fiber systems	Noncontinuous	0.13	0.87	60,000	15,000	24.7%

* $F_{Theoretical} = A_f A_f + A_m F_m$: where $E_f = 400,000$ psi; $F_m = 10,000$ psi. with NOL (Naval Ordnance Laboratory) type specimen F is 200,000 psi
† Typical test data
‡ Composite efficiency equals test stregth of composite/simple theoretical composite strength; see Eq. 15.64

Table 15.63 Composite efficiency of RPs

Loading	Beam ends	Deflections at	K_m	K_s
Uniformly distributed	Both simply supported	Midspan	5/384	1/8
Uniformly distributed	Both clamped	Midspan	1/384	1/8
Concentrated at midspan	Both simply supported	Midspan	1/48	1/4
Concentrated at midspan	Both clamped	Midspan	1/192	1/4
Concentrated at outer quarter points	Both simply supported	Midspan	11/768	1/8
Concentrated at outer quarter points	Both simply supported	Load point	1/96	1/8
Uniformly distributed	Cantilever, 1 free, 1 clamped	Free end	1/8	1/2
Concentrated at free end	Cantilever, 1 free, 1 clamped	Free end	1/3	1

Table 15.64 Examples of loading conditions

Chapter 16
Other Processes

INTRODUCTION

There are major families of processes that each process different amounts of plastic in the United States and worldwide. They are extrusion, which consumes approximately 36 wt% of all plastics; injection molding (IM), which consumes 32 wt%; blow molding, which consumes 10 wt%; calendering, which consumes 8 wt%; coating, which consumes 5 wt%; compression molding, which consumes 3 wt%; and others, which collectively consume 3 wt%. Thermoforming is the fourth major process used; it consumes at least 30% of the extruded sheet and film as well as other plastic forms whose principal uses are in packaging.

When analyzing the processes to produce all types of products, at least 65 wt% of all plastics requires some type of specialized compounding. They principally go through compounding extruders, usually twin-screw extruders, before going through equipment such as injection-molding machines, extruders, and blow-molding machines to produce products.

As reviewed in chapter 3, there are over 600 different processes (Table 3.2). During the last century, many designers, researchers, engineers, chemists, fabricators, material suppliers, equipment suppliers, and others have been able to manipulate the basic temperature, pressure, and time fabricating plastic cycle to their advantage. For example, new materials may require certain processing techniques, with most of them eventually modified and used in the more popular processes.

Plastics compete with other materials (steel, aluminum, wood, glass, etc.). There is also extensive competition between different types of plastics and different plastic-manufacturing processes. All of this activity helps to expand the development of new plastic materials and new plastic processes.

Many of these processes overlap as to how they operate, and most meet the specific needs to produce a specific product. Unfortunately many of these new processes reinvent the wheel: they may add a decorative surface that is important but not earth-shattering. An example of overlapping is the so-called reinforcing plastic (RP) processes. When this part of the plastic industry developed over a half century ago, thermoset (TS) polyester-glass fiber RPs were bag molded, autoclaved, filament wound, and so on (chapter 15). They expanded in using other materials, reinforcements, and processes that include all the major processes and a few more. For example, we have had the process of reinforced IM (RIM); in fact more than 50% of all RPs go through RIM machines. Other examples are casting, encapsulation, and potting. These terms are often interchangeable; they interrelate very closely to describe slightly different processing techniques.

Out of this experience come new processes, such as transfer compression molding and the more recent reaction IM (RRIM). Very important is the fact that these developments continue to advance the use of the basic processes used in the industry. Those basic processes and a few others have been reviewed in this book, but this chapter reviews more important factors. and details a few of these other processes. Some of these processes serve important markets and will interrelate with the processes already reviewed.

PVC PLASTISOL

Introduction

This vinyl dispersion industry provides many different products worldwide and uses many different processes: slush molding; dip, rotational, spray, and continuous coating; and open, closed, and dip molding. Vinyl dispersions come in basically two types, known as plastisols and organosols. Plastisols, the main type used, are mixtures of polyvinyl chloride (PVC) plastic and plasticizers that, with the application of heat, can be processed by different methods—casting, coating, dipping, rotational molding, and spreading—to produce either flexible or hard parts.

PVC plastisol is a liquid suspension of a finely divided plastic (about 1 μm) in a plasticizer. With heat, the plasticizer is absorbed into and solvates the particles so that they fuse together to produce a homogeneous plastic mass. Many different products are fabricated with this method, including toys, beach balls, squeeze syringes, gloves, and interior parts for vehicles (110, 127, 143, 525).

PVCs are manufactured either by suspension polymerization or dispersion polymerization (105). Dispersion PVCs are characterized by 0.1 to 0.2 μm-sized particles. The liquid or paste plastisol is manufactured by suspending the dispersion resin in a plasticizer as shown in Figure 16.1, resulting in fluid suspensions of special, fine particle size in plasticizing liquids. When the PVC is heated, fusion or mutual solubilization of the plastic and plasticizer occurs. The dispersion gel turns into a homogeneous hot melt. When the melted plastic cools, it becomes a tough vinyl product.

When the plastisol is heated, it passes through several characteristic changes. As the PVC approaches its glass transition temperature, the plasticizer begins to swell the PVC particles. The plastisol gels when the PVC has absorbed all the plasticizer, which takes place at a temperature about that of the PVC glass transition temperature (T_g; chapter 1). At this state, it is dry and crumbly and

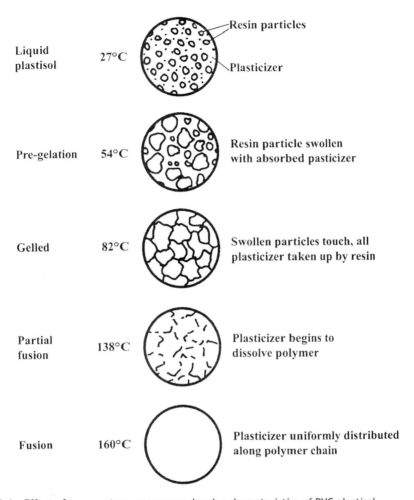

Figure 16.1 Effect of temperature on macromolecular characteristics of PVC plastisol.

lacks cohesive strength. Fusion and the development of physical properties begin when the plastisol temperature reaches approximately 280°F (120°C). By the time the plastisol temperature is approximately 380°F (190°C), the plastisol is fully fused but still liquid. Fusion is defined as the condition where the microcrystallites of PVC have fully melted and the plasticizer is fully dispersed through the PVC. The torque rheometer is the traditional testing tool for determining gelation and fusion conditions (chapter 22). A typical PVC plastisol isothermal time-dependent viscosity plot is shown in Figure 16.2. Although PVC plastisol is not a reactive polymer, it undergoes characteristic changes that relate to reactivity.

PVC has been known since the 1800s as a brittle, intractable, amorphous polymer that has very poor thermal stability in the presence of oxygen. It can be produced in crystalline form but all

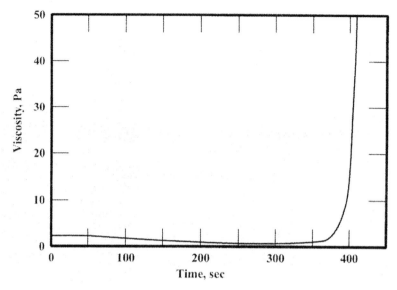

Figure 16.2 Example of time-dependent viscosity of PVC plastisol.

commercial grades are amorphous. PVC's chemical structure is $-(-CH_2-CHCl-)_x-$. In the early 1920s, Waldo Semon at BFGoodrich found that the PVC molecule could be solvated by many organics, particularly phthalates and phosphates. In addition, heat stabilizers based on heavy metals and on zinc and tin were developed to provide increasing processing life for the polymer.

To meet specific needs, other additives, such as lubricants, extenders, fillers, impact modifiers, and pigments, are added to the PVC compound in addition to heat stabilizers and plasticizers. Today it is estimated that more than 60% of all the adducts used in plastics are used in PVC compounds. Although the earliest PVC compounds were produced as emulsions, essentially all PVC compounds produced today are suspensions. Suspension compounds contain essentially no emulsifiers and are more processable. Liquid PVC compounds are called plastisols and typically have room-temperature viscosities of less than 10000 cp. Products made from plastisols are usually very soft. They have Shore durometers between 55A and 30A, and they can have characteristic skin- or leather-like appearances and feels (chapter 22).

Total world consumption of all types of PVC is at about 61 billion pounds. About 12% of PVC is used to produce plastisols (vinyl dispersions).

Processing Plastisol

The term *plastisol* is used to describe a vinyl dispersion that contains no volatile thinners or diluents. Plastisols usually contain stabilizers, fillers, and pigments along with the essential dispersion plastic and the liquid plasticizer. All ingredients exhibit very low volatility under the conditions for processing and use. Plastisols can be made into thick, fused sections with no concern for solvent or water blistering, as is the case in solution or latex systems, so they are described as being 100% solids materials.

When these 100% solids plastisols are formed, there is no significant shrinkage or increase in density. A small amount of shrinkage occurs because the vinyl compound usually shrinks more on cooling than the mold material does. This shrinkage is usually 1% to 6%, with the average at about 3%. Softer plastisols have higher shrinkage values.

The rheological behavior of the materials is important during processing. Their viscoelastic behaviors are complex. Plastisols may be shear thinning or shear thickening, depending mostly on PVC particle size, size distribution, and shape, but also on the type of plasticizer and other additives used.

Vinyl plastisols do have some disadvantageous characteristics. Dispersion grades of vinyl plastic, required to form the suspension, cost a little more than the more common general-purpose plastics. Process times are slow, usually 4 to 20 minutes in cycle time. Processing time can be minimized by using several low-cost molds when the production rate permits it. Another factor could be that the plastisol plastics contain wetting agents and soaps from the original polymerization reaction. These additives can cause clarity problems and can limit electrical resistance.

Vinyl, typical of plastics in that it is a good insulator, transmits heat slowly. The thickness of the finished product is determined by the amount of heat transmitted through the vinyl dispersion. Preheating temperatures tend to be high for thin molds and lower for thick molds. The finished product's thickness is also influenced by the length of time the mold is in the plastisol. (Chapter 17 provides information on molds.) The mold should be withdrawn from the plastisol slowly and smoothly to avoid the formation of lines on the product. With excessive draining, there will be runs and streaks on the product. The mold should ideally contain sufficient residual heat and should be withdrawn slowly enough that the fluid plastisol runs off and the remainder gels immediately without running or dripping. When this cannot be accomplished, it may be possible to rack the products so that runs drip from one corner; then, by inverting the products, the last drip can flow back.

Many different processing methods are used, all with heat, the essential element. These processes create products that would otherwise require costly and heavy melt processing and complicated molds and equipment. Different types of dispensing equipment are used to meet different flow rates and delivery amounts. No pressure or mixing is necessary. This means that mold costs are very low and the overall processing equipment costs are low. They are very versatile materials in that almost any additive can be incorporated for special effects as long as it is soluble in the plasticizer or can be ground to a powder sufficiently fine to be suspended in the plastisol.

A metallic mold is first heated to around 350°F (177°C) for a few minutes, which in turn heats the plastisol. The plasticizer is absorbed into the particles and solvates them so they fuse together to produce a homogeneous mass. The fusion process is called gelation. Plastisols and the related organosols, plastigels, and rigisols provide a liquid form of PVC to which special processing techniques may be applied. They are often more convenient for producing useful products than conventional melt processing methods reviewed in the other chapters of this book.

During processing, the plastisol is heated slowly. The first change that occurs is a slight lowering of viscosity. At a temperature of 120°F to 200°F (49°C to 93°C), the viscosity increases rapidly; this is the gel point or gel range of a specific plastisol. Different compounded plastisols have different gel points.

When the gel point is reached, the plastic absorbs the plasticizer. However, in a very soft compound, the plastic dissolves into the plasticizer. Because each plastic particle remains a separate particle, the resultant gel has no useful physical properties. But on further heating to 350°F (177°C), the plasticized plastic partially melts and flows into the plasticizer; this occurs at the fusion point or over the fusion range. On cooling, the material comprises the tough rubber compound known as a flexible vinyl.

As the temperature is raised, the viscosity of the plastisol changes, starting at a low viscosity, increasing over the gel range and peaking at the onset of fusion. The viscosity goes down during fusion. Satisfactory processing of vinyl plastisols requires an understanding of gelation and fusion, their mechanisms, and their effects on molding and cooling. At 350°F (117°C) the plastisol possesses practically 95% of its final physical properties.

Because the vinyl is a relatively good insulator, it takes time for the heat to penetrate completely. It is often advantageous to set the oven at 375°F to 400°F (191°C to 204°C) to shorten the fusion time. Care must be taken at high temperatures to avoid exceeding the heat stability of the compound and causing it to degrade.

Processing Organosol

It is sometimes convenient to extend the liquid phase of a vinyl dispersion with organic volatiles that are removed during fusion. The term *organosol* applies to these dispersions. An organosol is a suspension of a finely divided vinyl plastic in a plasticizer (diluent) with a volatile organic liquid. The plastic does not dissolve appreciably in the liquid at room temperature but does at elevated temperatures at which the liquid evaporates. Upon cooling, a homogeneous plastic mass is produced.

Organosols are similar to plastisols in that part of the plasticizer is replaced with a solvent. These less expensive vinyl dispersions were very popular in the past, before legislative action limited the amounts of solvent that processors could discharge into the atmosphere. They are still used, but now they must have safe ventilation systems (an added cost for the processor).

Organosols are suspensions of finely divided plastic in a volatile organic liquid. PVC is most frequently used. Plastic does not dissolve appreciably in the organic liquid at room temperature, but it does dissolve at elevated temperatures. The liquid evaporates at an elevated temperature; the remaining residue upon cooling is a homogeneous plastic mass. Plasticizers may be dissolved in the volatile liquid. Organosol production is more cost-effective with a solvent-recovery system.

Slush Molding

Slush molding produces hollow products from vinyl plastisols. It is the reverse of plastisol dip molding and an offshoot of open molding. Slush molding has been extensively used for making dolls, balls, flexible toys, hip boots for fishing, automobile parts (gearshift boots, armrests, headrests, etc.), safety cones, and others products.

The mold may be split or one piece. The finished part is removed either by splitting the mold or, in the case of a one-piece mold, by collapsing the part with a vacuum. This process can be very laborious. However, it is also automated and requires almost no manual labor. Automatic systems fill molds with plastisol carried by conveyor belts through an oven as the plastisol is being slushed (the mold is put into a control-motion pattern). The plastisol can gel repeatedly to a thickness of 0.06 in (15.2 mm). The excess plastisol is poured out of the mold and automatically returned to the main tank for reprocessing. The molds proceed to another oven, where curing is completed.

The process is dependent on temperature and time. It goes through the following stages:

1. A mold with a female cavity is preheated.
2. A mold cavity is filled with a measured amount of plastisol.
3. The plastisol in the heated mold is dispersed (slushed) evenly over the inner cavity surface by the back-and-forth motion of the mold, side-to-side motion, or rotation or all three, usually around one axis (usually the vertical axis).
4. The heated mold causes gel to form.
5. Excess plastisol is drained out of the mold.
6. The plastisol is heated to fuse.
7. The mold and plastic are cooled.
8. The product is stripped from the mold and trimmed if necessary.

The mold is preheated sufficiently to gel the required thickness of plastisol. The mold is filled and held for several seconds before it is inverted and drained. Another technique is to fill a cold mold then heat it to gel a skin on the mold cavity. This action can improve the reproducibility of the mold texture, a grain, or an engraving on the cavity wall. This action will be reproduced on the cured product with either a preheated or cold mold. Care is required when using a cold-mold approach because the plastisol can gel on the air in the cavity or its outside surface, producing poor drainage and forming lumps in the plastisol. The lumps can cause the product to have uneven thickness or performance or both. The lumps have to be screened out of the remaining plastisol or they will redeposit (or recycle) on the following slush moldings, causing further problems.

ROTATIONAL MOLDING

The rotational molding of vinyl plastisols is similar to the rotational molding of powder resins (chapter 13). A measured amount of plastisol is placed in the mold and heated while the mold is rotated. Molded products made with this method include tanks, volleyballs, basketballs, doll heads and bodies, and various automobile parts.

Spray Molding

Plastisols in liquid form can be sprayed on molds or parts. The viscosity of these special compounds is nonflowing after they are sprayed. Thicknesses of up to 50 mil (1.3 mm) can be obtained in a single pass on a vertical panel. The sprayed parts are heated and cooled. Multiple passes can be made to produce thicker parts. After cooling they are stripped off the mold or left on as a coating. Many small to very large tanks are lined with sprayed plastisols or organosols.

With liquid organosols, it is possible to spray or cast a harder film coating than is feasible with plastisols. This is because the organosols solvent can produce a film with a thickness measuring 10 mil (0.25 mm) or less. Such films are used to replace paint films in special applications where the chemical resistance of vinyl is required.

Continuous Coating

This procedure identifies plastisols that are spread-coated on different substrates, such as paper, aluminum foil, and plastic sheets or films. Applications are made by a doctor blade, direct roll, or reverse roll operations. Examples include plastisols that are roll-coated on adhesively primed metal for house sidings and cloth fabrics saturated or impregnated and then coated in the manufacture of conveyor belts.

Other products, such as foamed vinyl fabrics, are also produced using continuous coating. One method of fabrication is to coat a thin layer of solid plastisol on embossed release paper, then coating a thicker coating of foam plastisol and finally layering on a cloth scrim. The composite is fused and peeled from release paper. The wear layer of vinyl flooring is usually a coated clear plastisol, making it a no-wax flooring.

Open Molding

Most of the plastisols are used in open molding. It is a very simple process. A measured amount of plastisol is poured into an open mold cavity. The mold and plastic are heated to gel and fuse the plastisol. The mold is then cooled so that it solidifies. It is stripped from the mold. Inserts can be placed in the liquid plastic before it is fused; inserts can also be placed in the mold before pouring. Two or more colors can be placed in different parts of the mold.

This process is used to produce all kinds of commercial and industrial products. Examples of these products include automotive air filters, tablecloths, coin mats, truck flaps, simulated worms and other baits for fishing, display items, various novelties, and many other relatively flat to complex products.

Applications with special compoundings include automotive oil filters, in which additives are included in the compound; the additives cause the plastisol to bond to the filter media and to the metal end cap. The plastisol then becomes both an adhesive and an end seal for the oil filter. Table 16.1 provides an example of a PVC formulation.

Ingredient	Common Type	Concentration (parts/hundred)
PVC suspension resin	Homopolymer, 0.68–0.74 IV	100.0
Tin stabilizer	Mercaptide, 13–20% tin	1.2–2.0
Processing aid	Methacrylate copolymer	1.5–3.0
Costabilizer/lubricant	Calcium stearate	0.5–2.0
Filler	Calcium carbonate, 1–3 µm	0–5
Pigment/UV stabilizer	Titanium dioxide	1–2
Impact modifier	ABS or MBS polymer	0–5
Lubricant	Paraffin wax or fatty acid amide or fatty acid esters	0.5–1.5

Table 16.1 Example of a PVC blend formulation

CLOSED MOLDING

This process is a takeoff of open molding, except it is closed like a two-part compression mold (chapter 14). A measured amount of plastisol is poured or pumped into the closed mold cavity, similar to closed molding except that a slight pressure of about 5 psi (34.5 kPa) is applied. The mold is heated to fuse the plastisol and then cooled. The mold is later opened and the product stripped out. This technique may provide for accurate thickness control, filling very complex surface configurations, and so on.

DIP MOLDING

Dip molding is a takeoff of dip coating. In dip molding, the solidified plastisol is stripped off the mandrel or mold. In dip coating, the vinyl and mandrel or mold become part of the finished product (chapter 10). The process for dip molding goes through the following stages:

1. A metal mold or mandrel is preheated in an oven.
2. A mold or mandrel is dipped in a tank of plastisol for a required period of time.
3. The coated mandrel or mold is removed from the tank.
4. Excess plastisol is drained off.
5. The mandrel or mold is returned to the oven and heated until the plastisol and the mold reach 350°F (177°C).
6. The part is removed from the oven and cooled to 130°F to 140°F (54°C to 60°C). The cooling can be done by hanging the product in cool air, spraying water on it, or by dipping it into water. (Care must be taken to avoid leaving watermarks on the plastic.)
7. The plastisol is stripped off the mandrel or mold while it is still soft enough to stretch and pull over undercuts but cool enough not to be distorted by stretching.

DIP COATING

Chapter 10 provides information on coating different types of plastics by different fabricating processes. This section pertains specifically to dip coating plastisols. The process for dip coating is the same as that of dip molding except that the mandrel or mold is part of the finished product. Plastisols do not adhere to metals or other mold materials. A primer adhesive is used if the coating requires adhesion. Primer adhesives are usually lacquers that may be dipped, sprayed, or brushed on the metal part before it is preheated. These lacquers are usually solvent-based or water-based adhesives (chapter 20).

Products made with dip molding are many. They include slip-on grips, medical gloves and instruments, automotive bumper guards and gear shift boots (accordion and straight boots), and electrical devices (transformer and car battery leads, bus-bar insulation tubes, electronic controls, etc.) It is also used for products produced in large quantities, such as tool handles that are insulated against heat and cold, insulated field coils for car starters, cushioned kitchen tools, electrically insulated devices that perform in hot or cold environments, and protective coverings for sharp tools.

HEATING SYSTEM

Ovens are a popular heating system for processing vinyl dispersions. They may be gas-fired or electric, convection or infrared. The key to their success is their uniform heat. It is important to provide sufficient exhaust to vent the smoke produced by the hot plastisol so that it is not exposed to the atmosphere. Any limitations of these ovens may reduce their economy and efficiency at producing the best-performing products.

Compared to other techniques, air is a poor medium for the transmission of heat. Other methods of heat transfer may be used as required. They include hot plates for open molding, electric resistance heating for preheating rods and mandrels, and hot baths of molten salt for slush molding. Molds may be corded for slush molding despite their very short cycle times. Cording may be accomplished by attaching coils or electric strip heaters to the mold and using hot oil for heating and cold oil for cooling. Infrared heats only what it sees.

INK SCREENING

A very popular application is to use plastisols for screening on T-shirts, sportswear, and other apparel (this process was previously called silk screening) The ink becomes heat-fused. These are highly pigmented systems with application thicknesses of only a few mils (~0.25 mm).

ENCAPSULATION

Encapsulation can also be called *embedding*. This process is the enclosing of products in plastic. These products are diverse, and they include solenoids, ornaments, medical devices, sensors, motor components, and integrated circuits. Different processes, from casting (chapter 11) to IM (chapter 4) can be used to encase products. Encapsulation is a simple process, similar to casting. It can be automated at a very low cost and used for high volume production. Inserts of any size, shape, and number can be encapsulated. Either thermoplastics (TPs) or TS plastics, such as milled glass fibers, can be used with or without reinforcements to meet different performance requirements (chapter 2).

To insert a product, it can be immersed in a plastic liquid prior to the plastic hardening. The product could have fixed spider-type supports, retractable pins, or other features to support it when molten plastic is poured or injected around it. Another approach is to place the product on a layer of plastic that is partially polymerized in a mold cavity; this is followed by the application of a final layer that physically encloses it. For certain plastics and products, a vacuum system can be used if to eliminate air pockets or voids.

POTTING

Potting involves casting a plastic in a shell container that represents a mold cavity. A product could be embedded within the plastic. Potting is similar to encapsulation except that the shell is not separated from the finished product. It is an embedding technique in which the shell and plastic remains consolidated.

LIQUID INJECTION MOLDING

Liquid IM (LIM) is a variation of the reaction IM process (chapter 12). The major difference is in the manner in which the liquid components are mixed. In the LIM process, the entire shot is mixed in a chamber before injection into the mold rather than being continuously mixed and injected, as in the RIM process. LIM is used to mold smaller parts that are below the desired capacity of RIM. LIM also allows higher-viscosity and filled materials to be processed.

LIM offers automated, low-pressure processing of (usually) conventional liquid TPs or TSs and RPs; LIM has faster molding cycles, low labor costs, low capital investment, energy savings, and space savings. Because the pressures of injection are low—approximately 25 to 50 psi (172 to 345 kPa)—very fragile inserts can be molded, and mold wear is at a minimum. Some formulations for LIM may be molded at temperatures as low as 200°F (93°C), which permits the encapsulation of some heat-sensitive electronic components that do not lend themselves to encapsulation at conventional transfer molding temperatures of 300°F (149°C) or higher.

As shown in Figure 11.2, LIM employs two or more pumps to move the components of the liquid system (such as a catalyst and a plastic) to a mixing head before they are forced into a heated mold cavity. In some systems, screws or static mixers are used. Only a single pump is required for a one-part plastic system, but usually systems of two or more parts are used.

Equipment with control systems of varied sophistication is available to process all types of plastic systems. A very critical control involves precision mixing. If voids or gaseous by-products develop, a vacuum is used in the mold. Plastics such as polyester, silicones, polyurethanes, nylon, and acrylic are used. An example of LIM with silicones is the encapsulation of electrical and electronic devices.

A two-part TS plastic system can be used (chapter 1). One of the parts is a catalyzed TS mix and the other part is an uncatalyzed TS with or without reinforcements. The parts are separated in the cavity by a rigid or flexible hose. Curing action occurs in the cavity and products are molded. Depending on the catalyst (accelerator, etc.), the system cure may or may not require heat.

Vacuum-Assisted LIM

The vacuum-assisted liquid molding or vacuum-assisted resin transfer molding (VARTM) process has been used for the manufacture of large composite parts. In this process, a preform is placed in an open mold and a plastic vacuum bag is placed on top of the mold. A vacuum is created in the mold using a vacuum pump. A resin source is connected to the mold. As a vacuum is drawn through the mold, resin infuses into the preform.

VARTM is increasingly used in the manufacture of large parts with complex geometry, such as panels of all-composite buses, railroad cars, and armored-vehicle components. These parts are manufactured after the process has gone through a costly and time-consuming development cycle. This development cycle is empirical and experimental and requires a considerable amount of effort and expertise.

However, it is unclear whether the manufacturing process is efficient and cost-effective. The actual processing conditions may differ from that of the development cycle on a part-to-part basis. Hence there is a need for a scientific study of the manufacturing process and the development of objective process-efficiency criteria that will facilitate cost-effective manufacturing. The increasing complexity of components and new processes, such as coinjection resin transfer molding (coinjection resin-transfer molding [RTM]), will lead to an increasing need for the optimization of RTM-based processes. The filling and cure of the part can be simulated by several packages developed for this purpose, such as LIM software that has been developed at the University of Delaware (178).

IMPREGNATION

This is a specialized method of embedding, used mainly for electrical coils and transformers, in which a liquid plastic is forced into the interstices of the component. Trickle impregnation, a related process, uses reactive plastics with a low viscosity, first catalyzing them and then dripping them onto a transformer coil or similar device with small openings. Capillary action draws the liquid into the openings at a rate slow enough to allow the escape of air displaced by the liquid. When the device is fully impregnated, exposure to heat cures the plastic system.

CHEMICAL ETCHING

Chemical etching is the exposure of certain plastic surfaces to a solution of reactive chemical compounds. Solutions used in chemical etching are oxidizing chemicals, such as sulfuric and chromic acids, or metallic sodium in naphthalene and tetrahydrofuran solution. Such solutions are highly corrosive and thus require special handling and disposal procedures. Chemical etching causes a chemical surface change, such as oxidation, thereby improving surface wettability and increasing its critical surface tension. It may also remove some material, introducing a microroughness to the surface.

Chemical etching requires immersion of the part into a bath for a period of time, followed by rinsing and drying procedures. This process is more expensive than most other surface treatments, such as flame treatment; thus it is used only when other methods are not sufficiently effective. Fluoroplastics are often etched chemically because they do not respond to other treatments; acrylonitrile butadiene styrene (ABS) is usually etched for metallic plating.

TWIN-SCREW INJECTION MOLDING

Glass fibers are a common filler or reinforcement added to plastics in order to improve mechanical and physical properties of the raw material, especially the impact strength and stiffness. The traditional route to producing fiber reinforcement involves blending the fibers into raw polymer in a twin-screw extruder followed by pelletization (chapter 5). The pellets are then molded using an IM machine (IMM) to form the final parts (chapter 4). All of these steps cause fiber attrition.

Programs to reduce fiber attrition are common. For example, 4.5 mm glass fiber strands were blended into polypropylene using a twin-screw extruder and then IM with tensile testing bars. Fiber length and distribution studies were then conducted on the parts. Final glass fiber lengths of 0.47 mm (number average) and 0.7 mm (weight average) were reported (178). The important conclusions were as follows:

1. Most of the fiber damage occurs in the IMM, rather than in the compounder.
2. Low screw speeds and relatively high barrel temperatures minimize fiber breakage.
3. Lower back pressures and generous gate and runner dimensions are recommended to preserve the fiber aspect ratio during molding (see chapter 19).
4. Back pressure has a more dramatic effect on the fiber length than the injection speed. This is understandable because the pellets containing glass fibers have to go through the melting section in the molding machine, which is a region of high shear and high frictional forces.

The twin-screw IM extruder (TIME) is an IMM that is capable of both blending and extruding in one step. Because it is a one-step process, the fibers never go through the entire extrusion process as well as the pelletization, which limits the fiber size; the pellets do not go through the melting section in the TIME either, but they are blended into the molten plastic before injection. The screw part of this machine is based on a nonintermeshing, counterrotating twin-screw extruder (NITSE; chapter 5) but differs in that one of the screws is capable of axial movement and has a nonreturn valve on the end. This action enables the screw to inject and mold parts. The result of this single-step operation is a 30% glass fiber length improvement that in turn improved the properties of the molded part.

TEXTILE COVERED MOLDING

Decorated moldings can be produced on a standard IMM, using conventional IM or injection compression molding (ICM; Fig. 4.59). Both processes have some advantages compared to the decoration of moldings by adhesive bonding, because both are one-step processes that avoid adhesives and have a good reproducibility (178).

When producing decorated moldings directly in an injection mold, the decoration material is stressed by high temperatures and cavity pressures. This could lead to strong damage of the decoration material. For example, one major type of damage common to foam material is the collapse of the foam layer of the decoration materials. This foam layer provides a "soft touch" effect to give products a comfortable appearance. This effect is lost if the foam layer collapses.

ICM is a special process of IM that is able to reduce cavity pressure. This process starts with the closing of the mold until the compression gap is reached. At this point in time, the mold is already sealed because it is equipped with shear edges. In the next step the melt volume, which is needed to fill the molding, is injected into the cavity, and the shutoff nozzle is closed afterward. In the last step, the mold is closed completely by the compression movement. Due to this movement, the melt is spread throughout the cavity until it is filled.

Finally, the melt is compressed because the injected melt volume is usually higher than the cavity volume. This is necessary to compensate for the volume shrinkage. In conventional IM this is done by the packing phase, which is not used in ICM. With optimization control of the ICM process, this is an economical way to mold improved and quality decorated parts.

MELT COMPRESSION MOLDING

In-mold laminating (ILL) and in-line molding (ILM) have been in use for over a century in developing carriers, decorations, and other products. There has been an increased application of textile cover stock and leather substitutes, both preferably with a soft touch. This development was primarily initiated by the automotive industry to prepare for future trends, as is reviewed in Table 16.2.

The technologies described by the term *low-pressure IM* (chapter 4) can make substantial contributions to achieve these objectives. Meanwhile, other industries not connected with the automotive industry—furniture and packaging material manufacturing, for example—are applying the processes successfully. This is a trend gaining momentum through excellent results obtained by the

1. Growing demands for better and more comfortably appointed interiors of passenger cars and to a lesser extent of vans, busses, and trucks achievable.

2. The necessity of cost reduction i.e. by fewer manufacturing steps and less manual labor including finishing.

3. More safety by application of materials with higher impact and without splinters or sharp rupture lines after accidents as well as the use of foamed padding.

4. Ecological concerns to be overcome by lamination without adhesives, furthermore by composites suitable for recycling or uncritical incineration of waste or used parts.

5. Preservation of fossil energy by reduced vehicle weights also easing the strain on traffic surfaces as well as by substitution of processes heavy on energy like Azdel preparation and forming (Chapter 19).

6. A fair chance for agriculturally orientated economies replacing industrial fiber by regenerative fibers.

Table 16.2 Automotive industry objectives for decorative plastics

compression molding of melt strips of long glass fiber–reinforced TPs (LFTs) into technical, non-laminated parts.

As reported, low-pressure IM techniques have a lot in common to justify the general term. They are not a fundamentally new technology but the clever combination of known technical methods further developed and improved for the purpose.

Three techniques are steadily gaining importance and wider use, which indicates their outstanding capabilities. These innovative processes are (1) back injection, including the injection/compression molding; (2) melt flow compression molding; and (3) back compression by melt-strip deposition for in-mold lamination (IML) and the compression formation of fiber-reinforced TPs. Low-pressure IM techniques have a lot in common to justify the definition, as summarized in Table 16.3.

Mold design is a decisive factor for factors that determine molding success, such as dimensioning and location of the sprue gates, dimensioning of shear edges, flow aids, and cooling and ejector techniques.

BACK INJECTION

Back injection seems to be the most descriptive and probably the most popular of this process's many designations. The process is performed on conventional, mainly horizontal IMMs or, increas-

1. Predominantly hydraulic clamping units, vertical or horizontal are applied, modified from clamping unit for conventional injection molding.

2. Plastication occurs by means of a single screw extruder.

3. The melt is injected into a mold, closed or open, by a conventional injection unit adapted for high plastication and injection rates.

4. All low pressure injection molding processes are capable of in-mold lamination (IML) of decorative cover stock.

5. Part forming is performed at low internal mold pressure originally not exceeding approximately 100 bar (1450 psi); also established as the borderline for economical in-mold lamination (IML) with about the maximum sustained by cover stock materials. With the advent of LFT (long fiber reinforced thermoplastics) compression molding internal mold pressures up to 200 bar (2900 psi) are applied that is an acceptable demarcation line between low and high pressure injection molding. Generally speaking the internal mold pressure for the low pressure technologies amounts to 15 to 60% of high pressure applications.

6. Most development efforts are dedicated to the reduction and limitation of internal mold pressure during the forming cycle. These are areas influenced by the machines and the pertinent software that include melt injection profiles, pressure build-up and compression speed profiles, clamping force decompression profiles, reduction of flow length by sequential gate valve actuation (cascade valve control) or variable melt strip deposition.

Table 16.3 Definitions applicable to low-pressure decorating molding

ingly, on special machines with (relative to the clamping force) large mold-mounting areas and purpose-built injection units with high injection rates and low injection pressure.

The cover stock is inserted into an open mold. A shear edge mold permits draw-in of the cover stock during the closing cycle to avoid wrinkles and damage resulting from stretching of the fabric and yet is flash-tight during the injection. Preventing weakening joint lines (also a potential source of wrinkles) requires good melt penetration and elimination of any foam backs and/or special textile effects such as piles, plush finish, or leather grain embossing on foils. Injection occurs through carefully arranged gates with pneumatic needle shutoff nozzles, which are actuated for injection in a specific sequence called cascade control.

Molds for back injection are quite sophisticated. Apart from a complicated hot runner system incorporating the shutoff nozzles with their pertinent drives, all other mold elements, such as the ejector, core pulls, and slides, have to be accommodated in the injection side of the mold. Ejectors, core pulls, and slides are not acceptable on the decorative side (chapter 4).

A variant is the injection/compression cycle. During this cycle, and sometimes after a performing stroke for the cover stock takes place, the carrier material is injected into a partially open mold. The part is formed and laminated by the closing of the gap. The mold corresponds to a back-injection mold. The method has similarities with melt flow compression molding. Remarkable results can be obtained with back injection, provided the limitations of the technology, which are especially valid for larger parts, are recognized. The limitations are as follows:

1. A restricted influence on cover-stock preservation, fabric or foil, without a barrier back finish
2. A required back finish for the save process
3. A significant effect on foam layers
4. A lack of a soft touch
5. The risk of wrinkles or damage to the cover stock
6. Complications in the mold system, which may be heavy on maintenance

Melt Flow Compression Molding

Melt flow molding (the process has other names) is performed on vertical clamping units. The cover stock, perhaps preformed for deep parts, is inserted into an open mold. Then the mold is partially closed. The carrier stock is injected from below through a hot runner system and several generously dimensioned gates with pneumatically actuated needle shutoff nozzles (chapter 4). The melt available as cake around the gates is compression formed into the part by closing the remaining mold gap.

Shear-edge molds with hot runner systems, similar to those for back injection, are applied. Mold costs, especially for large parts, are probably the only drawback preventing a wider acceptance of the process. Another disadvantage may be the rather diffuse patent laws in some countries.

Melt flow compression molding equipment is similar to back-compression equipment, except that it is normally fitted with only one injection unit, which is permanently attached to the lateral inlet of the hot runner block (i.e., motion axes, as in back compression equipment, are not required). Back-compression machines equipped for melt flow compression molding are available.

Back Compression (Melt Compression Molding)

Melt compression molding (MCM) is a more general term covering both IML for the previous back-compression process and LIFT for all long fiber–reinforced applications. The term *back compression* is accepted for a process based on compression molding of a melt strip deposited in an open mold. Back compression is the process during which a cover-stock cutting is placed on a melt strip for simultaneous compression molding and lamination (IML) of parts. Melt-strip deposition also includes fiber-reinforced TP stock (LFT) with the subsequent compression molding of nonlaminated structural parts.

MCM-IML

Table 16.4 reviews a typical MCM-IML cycle.

TPs are used as carrier materials. Predominantly polypropylene (PP) unfilled or talc filled is used, and so too are ABS, ABS/polycarbonate (PC) alloys, and PA, though to a lesser extent. There is a vast variety of cover-stock materials that include woven and nonwoven fabrics with various finishes, such as pile and plush, and many colors, including sensitive dark blue. Barrier layers are the exception, even for fabrics as light as <200 g/m^2 (<0.7 oz/ft^2). Foils (TPO, PVC) with an embossed surface are increasingly used because of their excellent grain preservation.

The cover stock is normally placed on top of the melt strip in order to minimize the amount of time the melt is exposed to heat. With today's fast cycles, cover-stock insertion prior to melt-strip deposition is possible, except for materials with foam backs or foils without back protection (barrier layer).

Lower stock temperature, intensified mold cooling, and thinner walls have led to significantly shorter cycle intervals. Lower temperature and decreased internal mold pressures supported by ram decompression and a length-controlled retraction cycle are protecting the cover stock most efficiently. With cycle intervals around 60 seconds, MCM-IML reveals its versatility and economical feasibility.

1. The cycle starts with an open mould in a vertical press.
2. A horizontal injector equipped with a deposition head moves into the mold depositing a melt strip in the lower mold half during the retraction movement (x-axis).
3. A flat or preformed decorative cutting is placed on the melt strip.
4. The press closes molding the part by compression.
5. At the end of the cooling cycle the press opens for part demolding.

Table 16.4 Example of an MCM-IML molding cycle

Originally intended mainly for large-area parts (complete door trims, for example), there is a rapidly increasing interest in MCM-IML for the development of smaller parts. Since the introduction of MCM-IML for industrial production less than a decade ago, the process has seen significant progress in advantages and application engineering, as shown in Table 16.5. Industry-valid reasons for the acceptance of the MCM-IML process is reviewed in Table 16.6, while a few compelling reasons advising against the wide use of the process are reviewed in Table 16.7.

A comparison of costs can be made. The integrated MCM-IML method will result in cost savings of up to 40% over the conventional high-pressure molding and subsequent press-lamination processes. An average cost reduction of 20% to 25% is a realistic assumption.

PROCESSING COMPARISON

Tables 16.8 to 16.10 compare molding technologies. They serve both as a guide to applications and as a means of locating opportunities and gaps in the technologies. Technologies at similar locations in the tables can be considered as potential alternatives for each other (178).

In these tables, materials are divided into polymers (rubber and plastics) and nonpolymers (solids and gases). Solids, typically fibers, serve as reinforcements, whereas gases reduce density or increase stiffness for a given cross-sectional area by distributing material to increase the moment of inertia. The term *macroscopic* refers to dimensions that are on the order of the thin part dimension—a 2 mm diameter void in a 3 mm rod, for example. *Microscopic* means dimensions two orders of magnitude or more smaller (a 300-layer laminate) than the part thickness. The target for these processes is to achieve one-shot manufacturing without secondary fabricating operations. It requires an optimal combination of materials, processes, and geometry.

ADVANTAGES
- cost reduction (material and processing)
- simplification of process
- reduction of thermal stress on the PP polymer
- lower cost of investment
- floor space requirements
- more flexibility in regard to material analysis and characteristics, etc.

APPLICATIONS
- Appryl (Elf-Atochem) of France, brand name "Pryltex"
- Borelis of Finland, brand name "Nepol"
- DSM of the Netherlands, brand name "Stamylan"
- Ticona (Hoechst) of Germany, brand name "Compel"

Table 16.5 Examples of MCM-IML advantages and applications

1. Numerous degrees of freedom of parts design.

2. Molding of high/deep assembly fixtures including undercuts.

3. Elimination of 180° back wraps previously necessary for optical reasons.

4. Fabricate 90° back wraps with recess.

5. Sensitive treatment of delicate cover stock such as fabrics with pile and plush finish, embossed foils, also material with foam back.

6. Genuine uniform soft touch also on large area parts.

7. None or little protective back finish of decorative cover stock.

8. Lamination in one heat and cycle.

9. Absence of adhesives including those with aggressive solvents.

10. Secure intimate durable bonding of lamination.

11. Low content of manual labor.

12. Low internal stresses because of the compression cycle.

13. Most rejects, waste, and used parts are recyclable or suitable for incineration; no waste for classified disposal.

14. Lower cost of investment as with conventional process.

15. Uniform industrially reproducible high quality.

Table 16.6 Examples of valid reasons for using MCM-IML

1. Parts with areas too small for controlled deposition of the melt strip.

2. Significant undercuts at the fringes of the laminated area, and

3. Unavailability of space, even after tilting of the core, to securely deposit the melt strip.

Table 16.7 Examples of invalid reasons for using MCM-IML

Scale	Composition		
	Polymers combined with		
	Other polymers	Gas	Solids
Macroscopic	Blow Molding Coinjection Molding (MMP)* Multimaterial Molding (MMP) In-Mold Coating (MMP) Dual Molding (LPM)*	Gas-Assisted Molding Liquid-Gas Molding (LPM) Blow Molding	Laminate Molding (LPM)
Microscopic	Lamellar Molding	Microcellular Plastics Controlled Density (LPM)	Sheet Composites Reactive Liquid Molding

* Processes designated LPM (Low Pressure Molding) are described in Table A4; those designated MMP (Multimaterial Multiprocess) are given in Table 16.13

Table 16.8 Process and materials composition

Materials		Geometry			
		Macroscopic structure		Size & shape	
		Laminate	Segmented	Large	Hollow
Polymers		In-Mold Coating Coinjection Blow Molding Dual Molding (LPM)	Multimaterial Molding (MMP), Blow Molding	Sheet Composites In-Mold Coating (MMP) Injection-Compression (MMP)	Dual Molding (LPM) Blow Molding
Polymers and non-polymers	Solid	Laminate Molding (LPM) Reactive Liquid Molding		Laminate Molding (LPM) Sheet Composites Reactive Liquid Molding	
	Gas	Blow Molding Gas-Assisted Molding Liquid Gas Molding (LPM) Fusible Core		Blow Molding Controlled Density Molding (LPM)	Blow Molding Gas-Assisted Molding Liquid Gas Molding (LPM) Fusible Core

Table 16.9 Processing, materials, and geometry

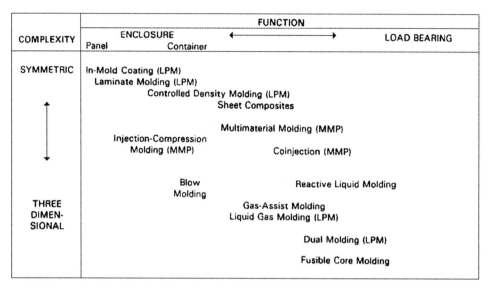

Table 16.10 Geometry function and complexity

Two of the innovative processes shown in Table 16.8, lamellar molding and microcellular plastics, serve primarily as the means of generating a unique material on the microscopic scale and secondarily for shaping the material. Other processes listed in the table combine polymers and nonpolymers on a microscopic or macroscopic scale to form products where specific properties of the multiple materials meet local functional needs. This localized optimization ultimately enhances overall product performance.

Processing is the bridge between the unshaped raw materials and the geometry (macroscopic structure and size and shape) of the product, as shown in Table 16.9. Many processes under microscopic structure allow the combination of a polymer with other polymer or nonpolymers on a microscopic scale to create laminates that have uniform properties over the surface of the part but variable properties through the thickness. In terms of part size and shape, innovative technologies are generally needed when the part is large, especially when a cosmetic surface is required, or when the part has a complex, often hollow shape, particularly when it is load bearing.

Part geometry can be considered in terms of function, where the part is located on a scale ranging from enclosure (containers or panels, often with cosmetic surfaces) to load bearing, and complexity, which allows for geometric complexity ranging from symmetric (planar or axisymmetric) to fully 3-D. Table 16.10 suggests that molded parts and the associated innovative processes to make them generally range from more or less symmetric enclosures (in the upper left) to 3-D load-bearing parts (in the lower right).

Tables 16.11 to 16.20 summarize a few processing technologies (178). They list advantages and disadvantages, applications, and materials.

Abbreviation	Adhesive
ABS	Acrylonitrile-butadiene-styrene copolymer
EVOH	Poly(ethylene-co-vinyl alcohol)
PA	Polyamide
PBT	Poly(butylene terephthalate)
PC	Polycarbonate
PE	Polyethylene
PEN	Poly(ethylene 2,6-nathalenedicarboxylate)
PET	Poly(ethylene terephthalate)
PMMA	Poly(methylmethacrylate)
PO	Polyolefin
PPS	Poly(phenylene sulfide)
PP	Polypropylene
PS	Polystyrene
PSU	Polysulfone
PPO	Polyphenyleneoxide
PPE	Poly(phenylene ether)
SAN	Poly(styrene-co-acrylonitrile)
SMC	Sheet Molding Compound
T-LCP	Thermotropic Liquid Crystal Polymer
TPU	Thermoplastic Polyurethane
TPE	Thermoplastic Elastomer

Table 16.11 Listing of abbreviations used in the following tables

Reactive Liquid Composite Molding (RLCM) proceeds in two steps: (1) PREFORM FORMATION by organizing loose fibers into a shaped preform, and (2) IMPREGNATION of the fibers with a low viscosity reacting liquid. The reacting material may be thermally activated by heat transfer in the mold or mixing activated by impingement of two reactive streams. Simulations of flow and reaction, a recent innovation in RLCM, allow determination of vent and weld line locations, fill times, and control of 'racetracking' in terms of gate locations, mat permeability, and processing conditions. Commercial success requires (1) fast reaction and (2) efficient preform formation. Cycle time for thermally active systems can be decreased by using higher mold temperatures and heating the preform. Innovative processes for PREFORMING include:

Thermoformable Mat heated by IR to melt the binder and pressed into shape by one or two moving platens while supported by a hold/slip edge clamp to reduce wrinkling.

Automated Directed Fiber Performers employ multiple delivery systems to create a surface veil, a chopped roving layer, and continuous roving with loops, all of which are fused by hot air.

The *SCRIMP* process channels resin flow between layers of fibers or along internal networks.

Water Slurry Deposition positions fibers by water flow through a contoured screen and sets them with hot air.

Innovations to reduce costs by combining process steps include:

Direct Part Forming combines sheet formation and shaping, e.g. heating porous sheet and then consolidating and shaping in a compression mold.

The *Hot Air Preformer* produces performs by either directed fiber or thermoplastic mat forming.

The *Cut-N-Shoot* process combines preforming and molding steps consecutively in the same tool.

Bladder inflation inside a mold shapes the preform and forms the mold wall during filling.

Advantages/Disadvantages	Applications	Materials
Low pressure and temperature processing by RLCM allow the use of inexpensive light-weight tools, especially for prototyping. RLCM allows customizing reinforcement to give desired local properties and part consolidation via complex 3D geometries.	Marine and Poolside Products, Sanitary ware, Caskets, Automotive Panels, Vehicle Suspension Links	Isocyanate based resins (mixing activated) Unsaturated polyester and styrene (thermal activated)

Table 16.12 Reactive liquid composite molding

The use of multiple materials and processes is the overarching technology in achieving one-hot manufacturing for large and/or complex parts. Material thermal expansion differences can be dealt with by flexible joints, (adhesives), process sequence to minimize distortion, sliding at interfaces (incompatible materials), and design for minimum distortion. Common multimaterial multiprocess technologies include Injection Compression Molding in which resin is injected into a partially open mold which closes, requiring less clamp force and producing less residual stress in the part. Multimaterial Molding in which a material is shaped, the mold is altered, and a second (or subsequent) material is shaped. Shaping processes are combinations of injection and compression molding and stamping. In-Mold Coating in which a thin thermoset coating is injected onto an injection or compression molded part in a closed mold. 'Mono-Sandwich' Coinjection Injection Molding in which a small extruder, operated intermittently, pumps a skin layer into the front of the main injection unit for subsequent coinjection. The Alpha 1 machine at GE Plastics, with two injection units, a long stroke vertical press, and shuttle table, allows combinations of compression molding, (gas-assisted) injection molding, and stamping. Other MMP technologies are described in tables on Low Pressure Molding, Advanced Blow Molding, and Lamellar Molding.

Advantages/Disadvantages	Applications	Materials
The advantages of more than one material and/or process include design flexibility, tailored performance, effective material use, lower labor costs, improved quality through automation, reduced secondary operations, less auxiliary equipment, and more recycle use. Multiple materials allow advantageous combinations such as multiple colors (automotive lens), flexible/rigid (conduits with connectors), and consolidated/strong (plastic/metal composite) and cost /barrier or strength (laminate structure).	Telephone booth molded from 132 lbs of structural foam on a 2500-ton press with three injection units Air vent with molded movable louvers made from incompatible materials. Automotive bumper with injection molded fascia over a stamped beam. Multi-color automotive taillights.	Combinations of thermoplastic, thermoset, and reinforcements subject to constraints of product performance, limitations on distortion, and interface requirements (adherent or incompatible)

Table 16.13 Multimaterial multiprocess (MMP) technology

Complex hollow parts are formed by injection molding plastic around a fusible alloy core which is subsequently removed by melting. The fusible (or lost) core typically is cast form a bismuth-tin alloy with a eutectic melting point of 138°C. The molten metal fills a split steel mold from the bottom and then cools for 2 min to produce a heavy core with a mirrorlike surface. The still-hot core is positioned by a robot in a steel mold and plastic is injected.

Flow channels are designed to balance forces around the core during filling to prevent core movement. For thermoplastics the injection temperature, e.g. 290°C for polyamide, can be well above the melting point of the core since the relatively high thermal diffusivity of the metal maintains a low interface temperature. After demolding, cores are melted out in a large bath or by induction heating or by injecting heat transfer fluid inside hollow cores.

Advantages/Disadvantages	Applications	Materials
Plastic parts made by fusible core technology have a weight and cost advantage over metal parts. Fusible core molding eliminates the need for mechanically complex molds or joining separately molded parts. Interior surfaces of fusible core parts are smooth which increases gas flow. Disadvantages are loss or oxidation of expensive core metal and need for robots to handle heavy cores.	Air intake manifolds, tennis racquets, pump parts	PA Poly(etherarylketone)

Table 16.14 Fusible core IM

Production of thermoplastic sheet composites involves two steps: (1) FORMATION of fiber reinforced sheets (prepreg) by polymer impregnation and sheet consolidation and (2) SHAPING of the sheets. Large volume competitive sheet FORMATION processes are continuous Melt Impregnation (e.g. Azdel sheet by extruding polypropylene onto a continuous "swirled' fiber glass mat), and Slurry Deposition in which long fibers and polymer powder with dispersing agents are deposited on a moving screen similar to paper making. Other processes are Powder Impregnation (powder and fiber consolidated by pultrusion, double belt press, or compression molding), Reactive Pultrusion, and Commingling (intertwining different fibers) and Coweaving polymer fibers, and reinforcing fibers. SHAPING techniques for consolidated prepreg include Melt-Phase Stamping (prepreg covering the mold cavity is heated with infrared and shaped in a fast closing press), Fast Compression Molding (thick charge flows during mold closing), and Solid State Stamping (semicrystalline plastics below their melting point are stamped into parts with simple geometries in 15 sec. Other shaping methods include Pultrusion of prepreg tapes, Diaphragm Molding (preform between plastically deformable diaphragms shaped by hard tooling), Rubber Pad Molding, Hydroforming (rubber bladder inflated hydraulically, Vacuum Forming in an autoclave, and Flexible Resin Transfer Molding (sheets of resin and fiber between elastomeric diaphragms are consolidated, then shaped).

Advantages/Disadvantages	Applications	Materials
Extrusion melt impregnation allows high fiber contents and longer fibers which give improved mechanical properties. Slurry deposition employs shorter fibers which allow greater flow and more complex parts. Cycle times are short.	Automotive Body Panels, Other Components, Aircraft Components	PP, PE, PA, PBT, PET, PVC, PC, PEEK, PSU, PPS

Table 16.15 TP sheet composite

PROCESS: Nitrogen gas under high pressure is injected through the nozzle or mold wall into plastic partially filling a mold. The gas flows preferentially through local thick sections with hot interiors and pushes the plastic ahead to fill the mold.

SIMULATION: Commercial software now available to predict gas flow paths, polymer thickness, clamp force, and contraction during cooling for various geometries and process variables including gas pressure, injection time, and prefilled polymer volume.

Simulations and experiment generally show

- increasing gas pressure decreases fill time, gas penetration distance, and (by conservation of mass) polymer wall thickness,
- melt temperature has a variable effect on gas penetration length,
- increasing delay time before the start of gas injection increases wall thickness and gas penetration length,
- increasing gas injection time increases gas penetration distance,
- decreasing the prefilled polymer volume fraction increases the penetration length until a critical level when gas blows through.
- increasing gas pressure level and time decreases shrinkage.

Simulations are generally able to predict undesirable air traps and gas penetration into thin sections. Simulation of a freezer bottom part converted to gas-assisted molding showed a 70% reduction in packing pressure, feasibility of using a less expensive material, and reduced warpage due to lower, more uniform pressure and higher part stiffness. Eight design guidelines are given based on both experiment and computer simulation [2.1].

Advantages/Disadvantages	Applications	Materials
PROCESS: Part weight and cooling time can be reduced up to 50%. Sink marks are eliminated. Warpage is reduced. Clamp force and injection pressure are lower. Part stiffness is increased because of the higher moment of inertia. Licensing is necessary. SIMULATION: Simulation helps identify optimal process conditions including runner layout and size, and location and timing of gas introduction. Software is available but new.	Handles, Panels with Ribs, Appliance/machine Housings (TV benzels) Automotive Parts (Clutch Pedals, Mirror Housings)	ABS, PA, PE, PP, PS, PPO, PC, PBTP, PC/PBTP, SAN, TPE, TPU

Table 16.16 Gas-assisted IM: process and simulation

Low pressure molding, as developed by Siebolt Hettinga, enables a number of other molding technologies. In Low Pressure Molding (LPM) the mold cavity is filled at low speed through large gates with a controlled pressure profile in the shape of a broad inverted U. LPM has no packing stage and no cushion. The melt temperature profile is controlled by adjusting screw speed and flow resistance during plastication. LPM works better with low viscosity semicrystalline materials and is not suitable for thin-wall parts. Slow injection, lower melt and higher mold temperatures reduce residual stress to allow demolding at a higher temperature to maintain or reduce cycle times. For larger parts, low clamp force can be achieved using multiple valve gates with programmed opening [4.2]. Lower clamp force allows use of self clamping molds and multistation injectors. Laminate Molding involves molding plastic at low pressure directly behind textile, film, or metal. In Liquid Gas Injection Molding, a volatile liquid is injected at low pressure into the melt and then vaporizes to form hollow channels in the part. The liquid condenses and is absorbed in the part. Dual Molding, similar to Bayer's Multishell Molding, forms an integrated hollow part by overmolding at low pressure an assembly formed from separately molded parts. In Controlled Density Molding the mold is partially opened once a skin has formed to give a low density interior.

Advantages/Disadvantages	Applications	Materials
Substantial capital costs savings result from the use of presses with a lower camp force or self clamping molds. Laminate Molding saves on assembly and adhesive costs in fabric/plastic laminates.	*Low Pressure Molding*: Interior Vehicle Panels, Bumper Fascia. *Laminate Molding*: Fabric/Plastic Seats, Vehicle Trim Panels. *Liquid-Gas Assist Molding*: Large Chairs, Chair Bases. *Dual and Shell Molding*: Manifolds, Pump Bodies, Valves, and Fittings. *Low Density Molding*: Fittings, Electronic Enclosures, Table Tops.	Thermoplastics, especially polyolefins, thermosets

Table 16.17 Low-pressure molding

The advanced blow molding technologies described below have greatly extended the versatility and facilitated product design. Deep-Draw Double-Wall Molding employs a mold with four hinged slides and an advancing core which close in a programmed manner around a partially inflated parison to shape a deep draw part. Press Blow Molding is used to form panels between shallow male and female mold halves which press together certain sections and inflate other sections to form hollow stiffening ribs. Three Dimensional (3D) Blow Molding forms serpentine three-dimensional parts without excessive scrap by manipulating the parison and positioning it in a convoluted mold cavity. Positioning the parison can be accomplished by (1) translating in two directions the mold which is titled at an angle, (2) movement of the parison by robotic arms in a mold with multiple sections which close sequentially, and (3) guiding the parison through the mold by sucking air along the length of the mold. Multimaterial Blow Molding employs multiple materials sequentially along the part length, in layers over the part thickness, or on opposite sides of the parison. New material developments include molding of 0.3-in fiber reinforced materials and foam layers [5.2]. Computer simulation of blow molding has been developed by A.C. Technology, Ithaca, NY in cooperation with G.E.

Advantages/Disadvantages	Applications	Materials
Deep Draw Technology increases draw (depth-to-length) ratio from 0,3 to 0.7 and allows forming of parts with undercuts, ribs, and noncircular crosssections. Multimaterial applications allow soft surfaces on structural parts, flexible conduits with rigid connectors, or parts with opposite sides of different properties. 3D Blow Molding consolidates complex parts and enhances function.	Insulated containers with foam, Planters, Conduits, Air Ducts, Bumpers, Equipment Panels, Instrument Panels, Portable Toilets, Golf Cases, Arm Rests, Gasoline Filler Tubes, Gas Tanks.	PE, PS, PP, POA, ABS, PPE elastomers

Table 16.18 Advanced blow molding

Extremely small closed cells from 0.1 to 10 microns in diameter can be formed in most plastics by dissolving gas in the plastic, typically supercritical CO_2, and then rapidly reducing the pressure and increasing temperature in a controlled manner to cause homogeneous and likely heterogeneous nucleation and growth of gas bubbles. The bubbles are to be smaller than naturally occurring flaws in the polymer so mechanical properties are not compromised. Particles in PS(HI) can be sites for heterogeneous nucleation [9.2].

Short diffusion paths, elevated temperatures, and gases in the supercritical state are necessary to achieve the high diffusion rate and high gas concentration needed for commercial use. Extrusion with gas injection is an efficient process to saturate polymer with gas. Manufacturing issues are determining sequence of pressure, temperature, and shaping geometries to nucleate and form cells without disruption and to shape product without distortion. Microcellular technology is covered by several patents and is offered for licensing by Axiomatics, Woburn, MA

Advantages/Disadvantages	Applications	Materials
The extremely small bubbles give weight reductions of 10-90%, no reduction in specific mechanical properties, appearance of a solid opaque surface, and foaming of thin sections. Fatigue resistance is observed to increase. Environmental advantages are use of atmospheric gases and lower material use	Siding, Pipes, Aircraft Parts, Athletic Equipment, Machine Housings, Automotive Components, Food Containers, Artificial Paper, Thermal Insulation, Fibers for Apparel and Carpets	ABS, PE, PET, PMMAPS, PS(HI), PP, PUPVC, SMC, Fluoropolymers, Poly(methylpentene)

Table 16.19 Microcellular plastic: formation and shaping

In Lamellar Injection Molding (LIM), two (or three) materials are extruded separately and combined with a 3 (or 5)-layer feedblock with multipliers to form a melt stream with hundreds of layers. This stream is injection molded to from parts with an irregular lamination pattern. The third material may be an adhesive. The layer structure, as assessed by oxygen permeability, shows (1) undesirable high permeability when too few layers allow easy passage around barrier layers, (2) low permeability (300-fold reduction) at 60-600 layers equal to theoretical minimum for lamellar structure, and (3) increased permeability as extremely thin laminates break up to from discontinuous domains (blends). LIM technology is offered for licensing by the Dow Chemical Company		
Advantages/Disadvantages	**Applications**	**Materials**
Only machine modifications needed are addition of feedblock and multipliers. LIM does not require multiple channels or sequenced valving used in coinjection molding and can easily be applied to complex parts or multicavity molds. Parts can be molded with high barrier properties to gases and hydrocarbons at lower costs than monolayer materials. Scrap can be recycled by incorporation into the major component or by conventional methods since LIM materials are compatible. Optical clarity (reduced haze) is improved compared to blends because the ordered LIM morphology reduces light scattering. LIM structure, with sheetlike continuous component selected for specific properties (controlled thermal expansion, increased load bearing, and temperature resistance), offers distinct property enhancements compared to blends.	Structural Parts (dimensional stability, temperature/ chemical resistance) Housewares/Durables (clarity, temperature and solvent resistance) Containers for Food and Chemicals (gas. hydrocarbon barriers) Automotive Reservoirs (fluid/heat resistance)	PC/PET, PC/ PBT, PO/ad/ EVOH, PET/ PEN, PO/ad/ PA, PS/PA6, PC/TPU, TP/ T-LCP, filled/ unfilled, brittle/ductile, virgin/recycle

Table 16.20 Lamellar IM